CIVIC GARDEN CENTRE LIBRARY
3000187
Civic Garden Centre
Library
D0171181

Botany for Flower Arrangers

Botany for Flower Arrangers

JOHN TAMPION
&
JOAN REYNOLDS

DRAKE PUBLISHERS INC
NEW YORK

2914

Published in 1972 by
Drake Publishers Inc
381 Park Avenue South
New York, N.Y. 10016

ISBN 87749-184-4

Printed in Great Britain

Contents

Illustrations

PLATES

LINE DRAWINGS

LIST OF TABLES

Table

Acknowledgements

Our thanks are due to numerous friends who for some time have posed questions which we cannot answer adequately and to those who have given so freely of their valuable experience with plants.

Those illustrations which are not our own are acknowledged in the captions. We are also grateful for other offers which we were unable to accept.

We would like to thank Dr C. L. Duddington and the Editorial Director of Pelham Books for encouraging us to write this book, and Maureen Tampion and Joan Young for deciphering our ever-changing notes and making them into a good typescript.

Foreword

Ours is a book on flower arranging without any flower arrangements in it.

There are so many books dealing with the creative and artistic side which have been written by well-known authorities that our attempts at this aspect would be superfluous. We would not presume to tell flower arrangers about their art.

We are botanists, trained to work with living plants and, perhaps, to appreciate their beauty all the more because of this training.

Dealing with the variable plant and the variable place of decoration brings its problems. You, who could never create the same arrangements twice, would appreciate this.

We have tried to give some of the principles of botany relating to floral decorations and to answer some of the questions which come to the mind of the flower arranger.

We hope this book will be of use to those who have graduated from 'doing the flowers' and 'creating lines' to wanting to know the how and the why of plants used in arrangements.

JOAN REYNOLDS
JOHN TAMPION

CHAPTER ONE

The Growing Plant

A healthy growing plant is the first stage in creating any flower arrangement. Poor blooms or foliage can never be satisfactory. To understand flowers we must first examine the growing plant both as a whole and in its various parts. It is well known that plants consist of roots, stems and leaves but few people realise just how these differ from one another.

The most logical way to begin is probably with a seed. Dry seeds, although living, show no visible signs of life. Only when they have taken up water can we expect them to germinate. This introduces us to the first essential for life—water. Only when water is present can growth take place.

The first visible sign of germination, apart from swelling, is the putting out of a root. This serves several purposes. Firstly it penetrates between the soil particles and draws upon water held by the soil so that the seedling may continue to grow. To enable it to do this satisfactorily a root is capable of reacting to several external influences. The growth of main roots is normally directed towards the source of any gravitational forces, which under natural conditions means the centre of the earth. This also coincides with the tendencies to grow towards water and away from the light which are found in some roots.

The principle job of a root is to take up water and dissolved minerals from the soil. At first, the root is white and only slightly protected but gradually the older parts become covered with a protective corky layer. Although this helps prevent diseases from attacking the root, it also hinders water entering the older roots. The young growing roots often have their capacity to take up water increased by the development of root hairs in a zone a little way back from the tip. The older parts of the roots have well-developed channels along which the water can flow towards the shoot. Often the roots also serve as a store-house for food reserves manufactured in the shoot and drawn down to the roots along another series of channels. These conducting systems form the vascular tissue of the root.

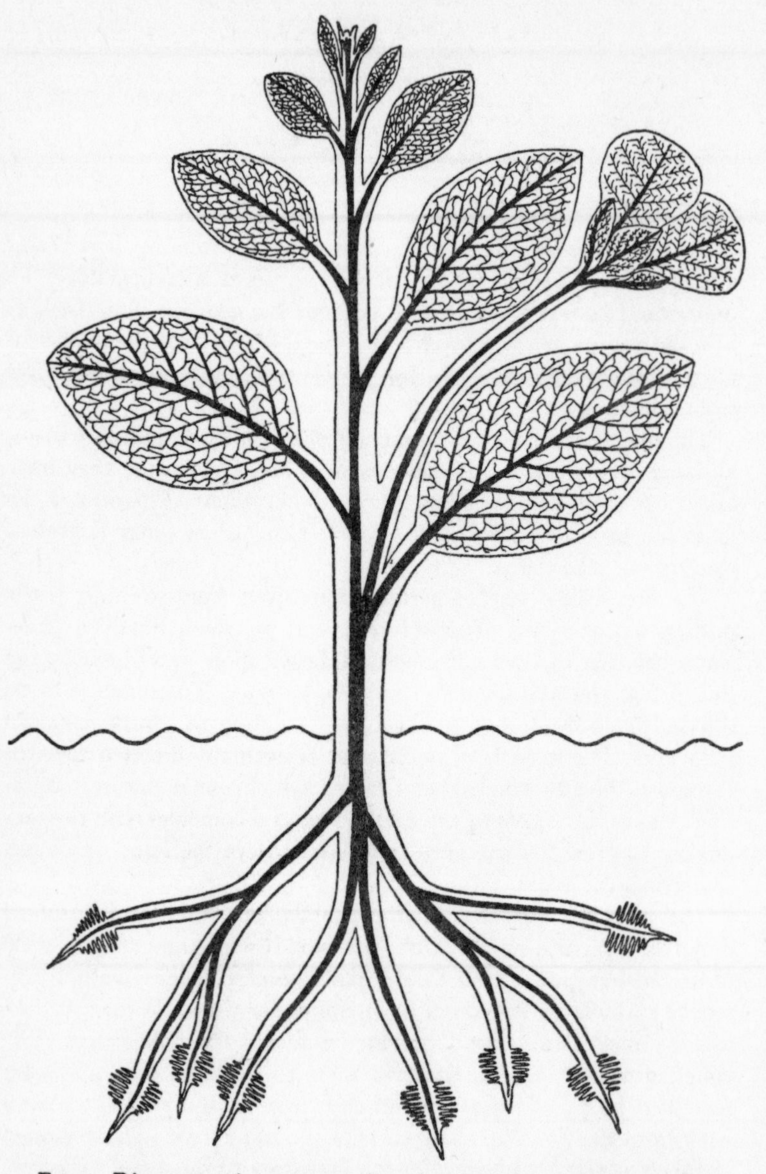

FIG. 1. Diagrammatic representation of a plant showing the vascular tissue extending from almost the tip of the roots to the tip of the shoots and out into the leaves and flowers, so that all the parts are interconnected

Vascular tissue consists of two parts—the xylem and the phloem. The xylem is mainly concerned with the flow of water and simple minerals absorbed from the soil, although in Spring the xylem sap may also contain some sugar derived from the reserves stored in the roots. The phloem sap consists of a much stronger solution of sugar plus small amounts of a wide range of other substances.

Under the microscope, plants are seen to be made up of various types of cell.

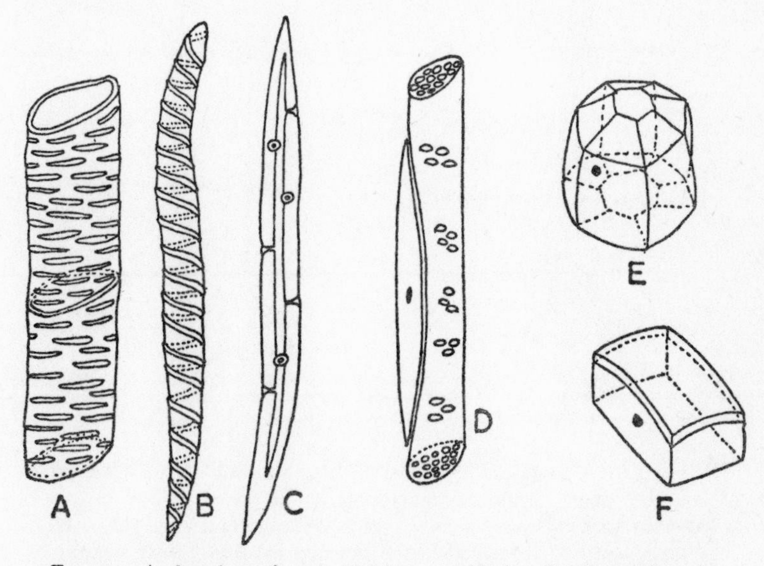

Fig. 2. A drawing of very highly magnified cells illustrating the wide range of shapes. (A) Vessel elements from xylem tissues. (B) Spirally thickened tracheid. (C) A fibre. (D) A sieve-tube element and companion cell from the phloem. (E) A thin-walled living parenchyma cell. (F) A flattened cell from the epidermis with characteristic cuticle on the outside wall

The main conducting cells of both the xylem and the phloem are elongated and may even link up into continuous systems of tubes with few cross-walls to obstruct the flow. The rest of the root consists of rather smaller cells which will often contain reserves of food. The cork cells which protect the outside of the older parts of the roots are thick-walled and waterproof.

The root system of a plant may consist of a single thick root with some much smaller ones growing from it or many smaller, fibrous, ones. In older plants the roots may become much enlarged with stores of food for over-wintering.

Now we must turn our attention to the shoot. Once again it is convenient to start with the seed. Botanists recognise two great groups of plants according to the number of leaves (or cotyledons)

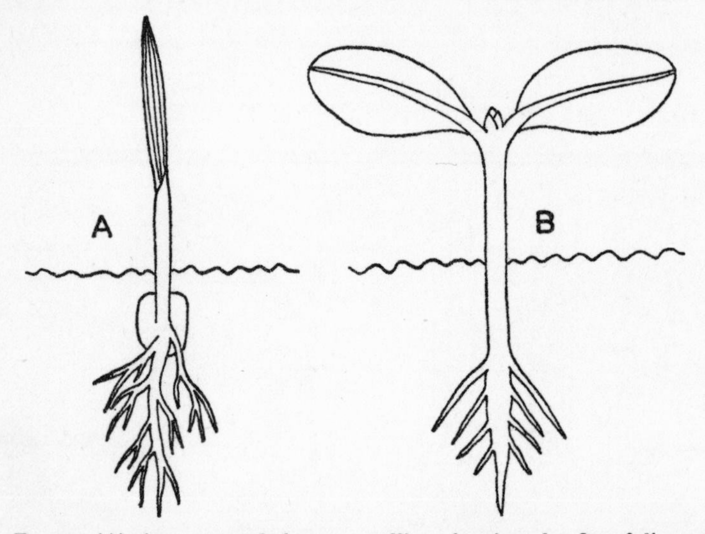

FIG. 3. (A) A monocotyledonous seedling showing the first foliage leaf; the single cotyledon remaining underground. (B) A dicotyledonous seedling showing two green, above ground cotyledons and an apical bud from which will develop the first foliage leaves

in the seed. The monocotyledon—monocots for short—have only one. Grasses and the multitude of flowers from bulbs and corms all belong to this group. An additional feature is the possession of long thin leaves with 'veins' which lie parallel to one another. Very few of these plants are woody and they usually have very little visible stem unless flowering. The other great group—the dicotyledons (dicots) have two seed leaves and their leaf veins are normally arranged like a net while the leaves may be almost any shape. To this group belong the familiar woody plants of the garden and field, roses, forsythia, oak and beech and the bulk of herbaceous garden flowers.

The main function of the shoot is to produce food for the plant.

This it does by using the energy contained in light. The green chlorophyll pigments are essential for this process. They absorb the light energy in the red and blue parts of sunlight, and hence, out of all the colours of the rainbow only the green is left for us to see. The light energy is mostly trapped in a substance known to botanists as ATP. The P stands for phosphate, and this is just one of the reasons why phosphate is so essential to plant growth. Some of the other light energy is used to break up a little (rather less than one percent) of the water taken up by the plant, and in doing so, oxygen gas, which is essential for animal life, is given out as a waste product.

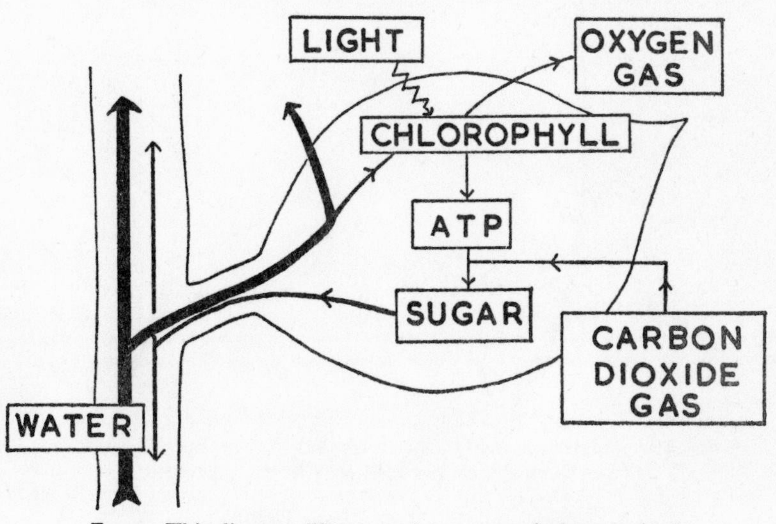

Fig. 4. This diagram illustrates the process of photosynthesis

The 'fixed' energy is used to help 'capture' carbon dioxide gas from the air and turn it into complex substances such as sugar and starch. The whole process is known as photosynthesis. It is the opposite of the reactions which go on in animals where carbon dioxide is given out and oxygen used up. Some scientists believe that all the oxygen of the earth has been liberated by photosynthesis. Others are even worried that by cutting down too many forests and destroying vegetation we may eventually have an excess of carbon dioxide in the atmosphere. Green plants, in the dark, use up oxygen just like animals and give out carbon dioxide. This is known as plant respiration and also occurs all the time in the non-green parts of plants.

Carbon dioxide, of course, is also produced when we burn up coal and oil. In normal plants most photosynthesis takes place in the leaves and the stems are often not green at all. We should be careful of saying that non-green plants cannot carry out photosynthesis for in some, such as the 'copper'-leaved cultivars often grown in gardens, the green pigments are present but masked by additional darker ones.

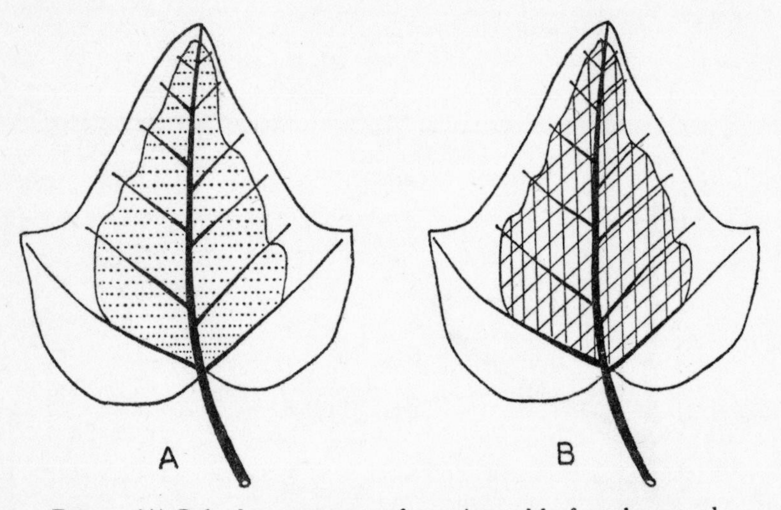

FIG. 5. (A) Only the green parts of a variegated leaf produce much starch by photosynthesis. (B) When treated to show up the presence of starch this is clearly seen only in the green parts

Some variegated leaves do not develop chlorophyll in the white areas and these must therefore draw their food from the green zones. Such plants are clearly at a disadvantage when in competition with fully green ones. 'Golden' cultivars contain much less chlorophyll than their green counterparts and are only about half as efficient at carrying out photosynthesis. Most variegated plants are therefore slower growing forms than their green relatives.

Under the microscope, it can be seen that leaves are quite complex and consist of several different kinds of cells. The bulk of the leaf is made up of rather soft cells which contain the chlorophyll. In the upper part of many ordinary leaves they are closely packed together like bricks stood on end, while in the lower half there are many air spaces between them. As we might expect, there is a complex vascular conducting system to channel water and minerals into the leaf and the products of photosynthesis out. These are the veins which we see in 'shadow' leaf skeletons. They may be quite woody in some plants, but in others have hardly any rigidity (see Plate 1).

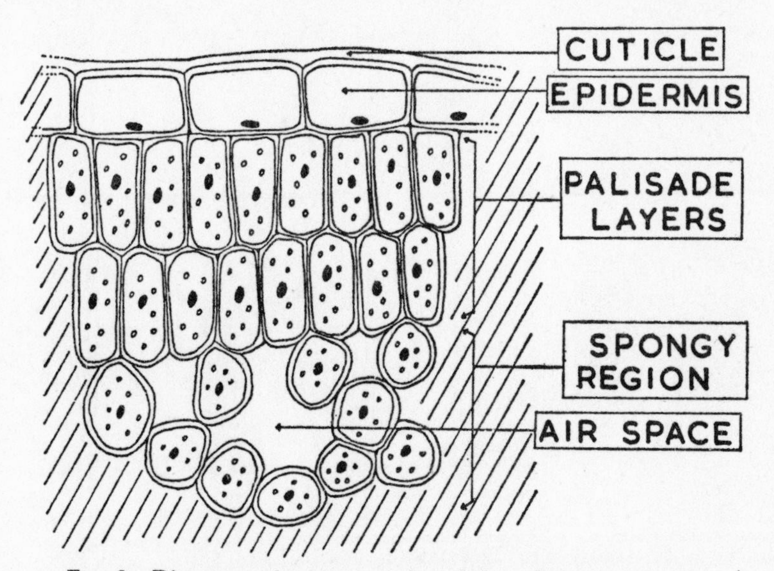

FIG. 6. Diagrammatic representation of the cells near the upper surface of a green leaf

A leaf, like all the rest of a plant, is about nine-tenths water and would very quickly dry out if it did not have some protective covering. This consists of a layer of cells called the epidermis. The outer surface of these cells is covered by a waxy, waterproof layer known as the cuticle. This markedly reduces water loss from the leaves, which have a large surface area, further increased by the air spaces in the leaf. Taking into the account the internal surfaces, an ordinary leaf has been estimated to have an actual surface area some six times greater than that which can be measured on the outside alone. Some plants, such as holly (*Ilex aquifolium*) have very thick waterproof coverings and can resist drying out for long periods. Others have only a thin covering and rapidly show signs of distress if placed in too dry an atmosphere.

Unfortunately, the waterproof covering tends to restrict the flow of oxygen and carbon dioxide into and out of the leaf, and so plants have tiny, controllable pores all over their leaves. These are known to botanists as stomata. Knowledge of their functioning is of great importance to the flower arranger (see Plate 2).

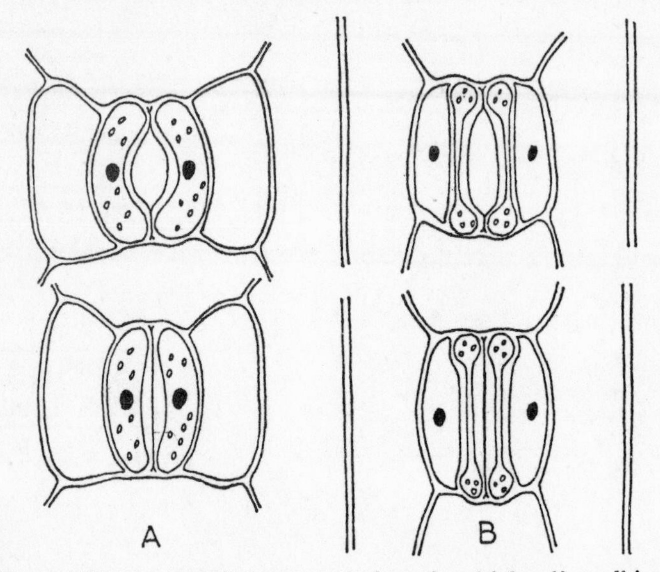

FIG. 7. Highly magnified stomata in 'open' and 'closed' condition. The diagram shows that stomata of dicotyledonous plants (A) and monocotyledonous plants (B) are typically different

The stomata are controlled in a very delicate way so that the plant can work at maximum efficiency. When the plant is turgid and well-supplied with water, the stomata will often be open but if only about five percent of the water in a leaf is evaporated away, they start to close up and this markedly slows down the further loss of water. Since their main function is probably to allow gases to pass in and out of the leaf, it is only to be expected that they should be open when carbon dioxide is in short supply. This occurs especially when a plant is in the light and 'fixing' carbon dioxide by photosynthesis. Often the stomata are closed when a plant is in the dark, except for many succulents which behave in the reverse and rather abnormal way.

The stem of a plant consists mostly of the conducting vascular tissues. Woody stems are mostly made up of xylem. Although many channels may be present, often only those produced in the current year are functional in water transport. The others are empty and serve only as a support for the stem and leaves. Outside the xylem is a thin zone known as the cambium. This produces new xylem elements to the inside and new phloem to the outside. If we want to kill a tree or shrub we 'ring-bark' it. This removes the phloem and cambium and prevents any food flowing down from the leaves to

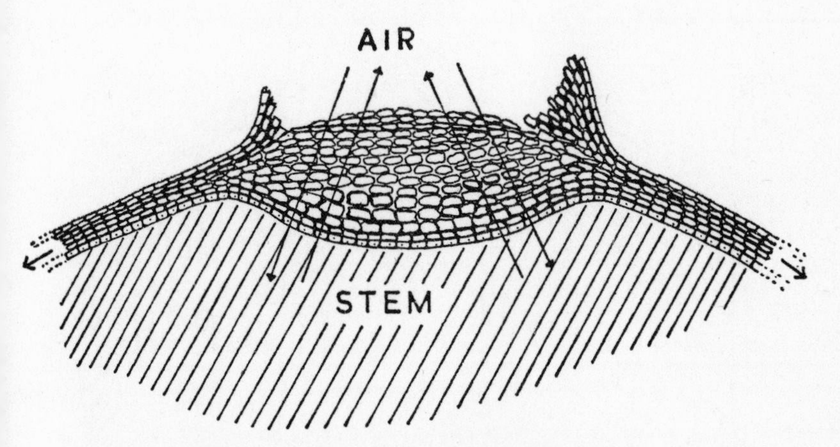

FIG. 8. Section of a lenticel showing loose corky cells exposed to the air. Lenticels form a passage through the cork covering the stems

the roots. Stems, depending on the type of plant and its age, are protected by anything from a thin epidermis to a thick corky layer (see Plate 3). Where a corky layer has developed, small holes often penetrate through it, presumably to allow gases to pass in and out. Botanists call them lenticels, and they are sometimes very obvious on woody twigs. Once again water can be lost from them but they have no mechanism which allows the plant to close them.

So far we have considered the various parts of a plant in isolation but in a growing plant they are all joined together into one system.

Consider first the vital question of water. This enters mostly by the roots since we have already seen that the shoot is almost covered with an impermeable layer. There is one vital principle we must now consider. Suppose we take a strong solution of any substance—say ordinary sugar—and place it inside a cellophane bag from which the sugar cannot escape but through which water can pass freely in and out. If the bag is now placed in water, we find that it will swell as water accumulates inside, tending to dilute the sugar solution. If there is nothing to stop it, the bag will go on growing larger. The

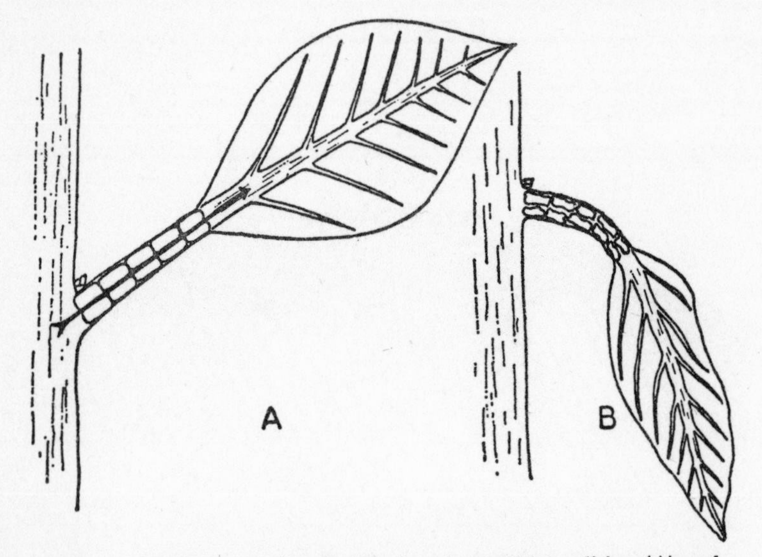

FIG. 9. Cells of a leaf stalk showing the turgid condition (A) and the flaccid condition (B)

tiny cells which make up a plant are very similar to this, but each one is surrounded by a fairly rigid cell-wall. When plentifully supplied with water, each one will eventually become firm and hard (turgid is the botanical term), at the same time the whole plant will also be firm. When water is in short supply the cells are flaccid and the plant will droop.

Some plants, such as willow-herb (*Epilobium angustifolium*) and the young shoots of trees, are supported by their turgor alone. These are the ones which show signs of wilting very quickly. They are very difficult material for the flower arranger. Others have quite thick cell-walls which are rigid even when the cell is not swollen with water. These plants are the stiff, woody ones which can stand up to rather harsher treatment. In many evergreens the covering of the leaves and the rigid veins are often strong enough to support the more delicate tissues inside and between them when they are short of water.

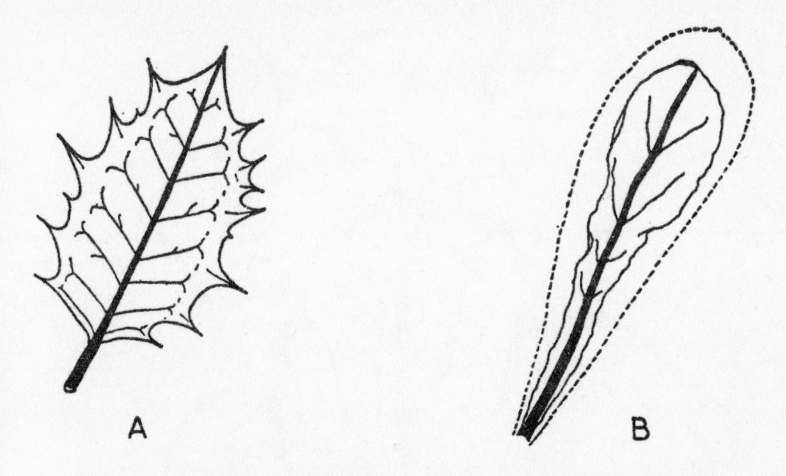

FIG. 10. (A) Holly leaf having rigid veins and tough cuticle keeps its shape at all times. (B) leaf of Marigold (*Calendula officinalis*) having weak veins and a thin cuticle shrivels quickly from its original shape (denoted by dotted outline)

The uptake of water is caused by the strong solutions inside the cells and is known as osmosis. If the cells vary in the strength of the solutions inside them the water will tend to flow towards those which contain the strongest solution. In this way water can flow from the soil into the roots of a plant. Actually the stretching of the cell walls contribute to this movement of water but for the moment we can disregard this. At certain times of the year, particularly in the spring in woody plants, the forces of osmosis assist in causing the sap to rise, and if a cut is made into the stem at this time, a sugary liquid exudes. This is due to 'root pressure'.

Movement of water by osmosis tends to be rather slow and plants have an additional and much more rapid means of moving water. This makes use of a completely different principle. We all know that if we leave a moist piece of material in dry air the water will evaporate from it. It is possible for surprisingly high forces to be developed by this evaporation. We have already mentioned the tiny pores in the leaf surface, and it is through these that the water is mostly lost.

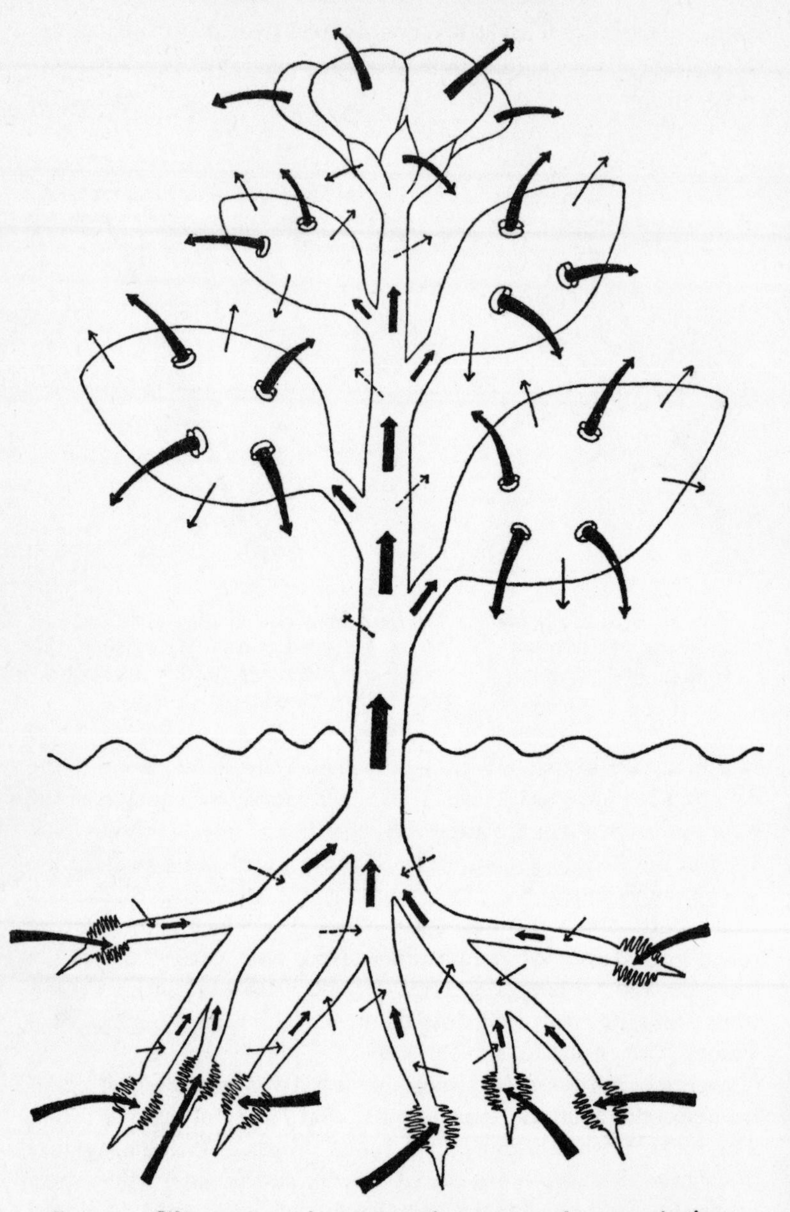

FIG. 11. Water enters the roots and moves as the transpiration stream up the stem and is finally lost as water vapour through the stomata and epidermis of the shoot

It is replaced by water moving up from the roots by way of the xylem of the vascular tissue. The loss of water from the plant is known as transpiration, and the flow of water in the plant as the transpiration stream.

We find that the water in the xylem is under considerable tension due to the forces of evaporation in the leaves. Actually the plant has

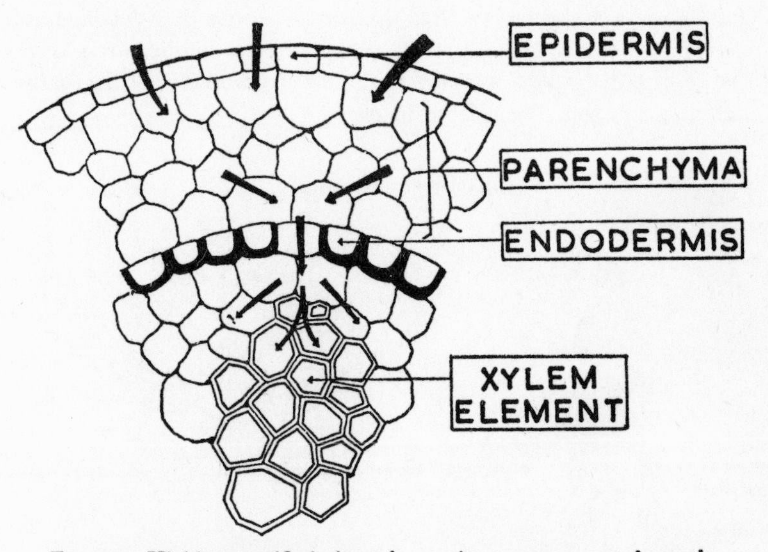

FIG. 12. Highly magnified view of a section across a root shows the endodermis which influences water movement into the xylem

a very extensive pathway for the water to flow in, since, not only the xylem, but also the cell walls of almost every other cell and any gaps between the cells, form one continuous system. One of the few blocks to the system occurs in the roots, where there is an almost complete cylinder of cells which have walls with a waxy, waterproof thickening to prevent the too rapid flow of water. Occasionally, so-called passage cells occur, which enable the water to flow more easily. The cylinder of cells is called the endodermis.

Contrary to what we might expect, roots often tend to hamper the flow of water, and for this reason, we sometimes find that water is more rapidly taken up by a shoot from which the roots have been cut off than from the whole plants with their roots in water.

An important consequence of the tension inside the xylem elements of a plant is that, when we cut across a stem, the water tends to shrink away from the cut ends—rather like a wire spring which has been stretched and then released. This is due to the fact that air cannot exist under tension—it just expands to fill the space—but water can, if no air is present. It should now be clear that as soon as we let air into the tiny channels of the xylem we are upsetting the flow of water. Of course, the water can still flow by other routes, but since these are slower, we find that air in a stem has detrimental effects upon the shoot. If the exposure to air has not been too long, the plant will usually recover, as the air is gradually displaced by water rising up the stem from below.

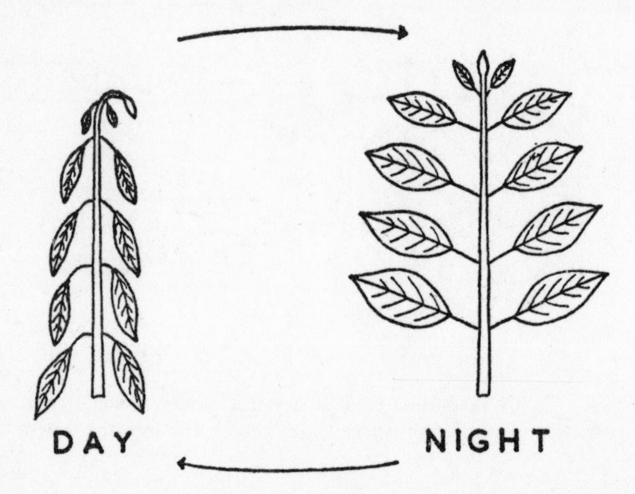

FIG. 13. The water content and turgidity of a plant often shows a daily rhythm, being greatest during the night. More water is lost during the heat of the day than the plant can absorb from the soil

We can now see the vital importance of the stomata to water flow. When they are open, evaporation can readily take place. In most plants, absorption by the roots tends to lag behind evaporation, so that we often find that a plant will wilt during the hottest part of the day and recover when conditions are cooler.

We all know, of course, that hot air can carry more water vapour than cold air—that is the reason for condensation on cold windows, or the morning dew. When the pores are closed, water loss is cut down. Usually the plant controls its own pores according to the amount of water in it, but there are certain chemicals which botanists are working with which can be used to close the stomata. They are useful in times of drought, or when transplanting, since many plants die, or are at least severely hampered, by excessive water loss. Most of the chemicals which have been tried have side effects on the plants which make them less useful than might be expected.

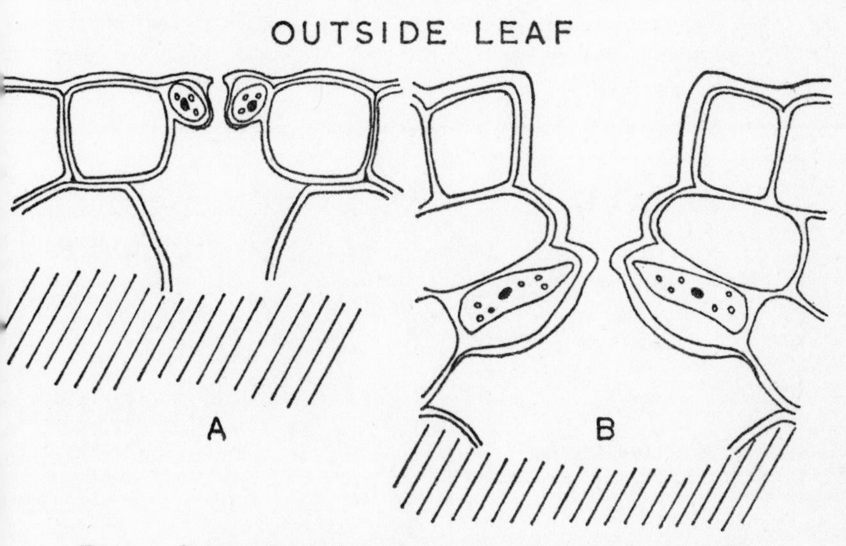

FIG. 14. Section of a leaf to show stomata (A) at surface and (B) sunken below surface

Naturally, the amount of water which is lost from a plant, depends on the surface which is exposed to the air. The larger the leaves and the more of them that there are, the faster the water can evaporate. It is partly for this reason that we often defoliate flowers such as Clarkia (*Clarkia elegans*), to reduce the surface area. Some plants have their own additional defences to prevent water loss. Leaves may be very small or even absent altogether, and in the case of certain grasses, they can even roll up to reduce the amount of leaf exposed to the atmosphere. Others again have their stomata sunk down in pits, or the leaf surface covered with hairs to cut down the flow of drying air.

Increasing movement of air naturally causes more evaporation to take place, because it sweeps away moisture which has been drawn out of the leaf. If the draught (or wind) is too strong, this eventually causes the stomata to close because of excessive water loss, but clearly, it is better to reduce the loss by standing cut flowers out of the draught.

Although our attention has been directed mostly towards the leaves, we should not imagine that they are the only organs from which water is lost. Flowers may have quite enormous surface areas, and since they often have only thin protective cuticles, they may be a major site of water loss.

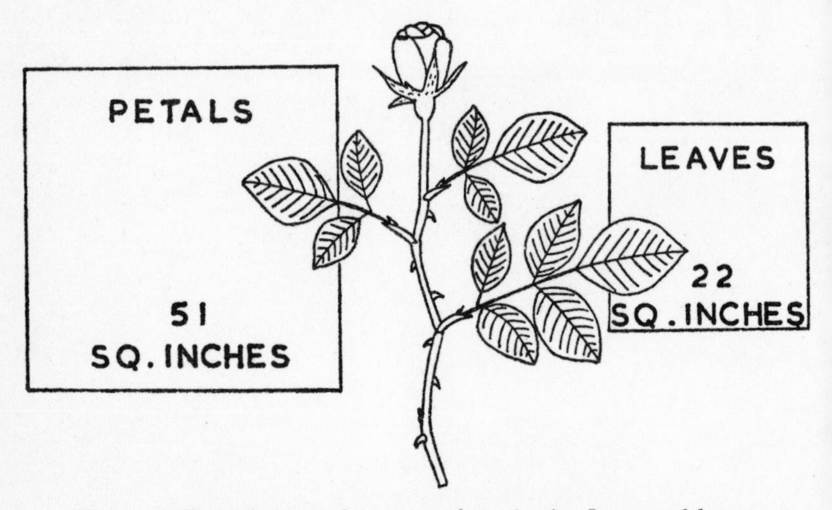

Fig. 15. The relative surface areas of petals of a flower and leaves on the flowering stem. Each petal and each leaflet was measured separately and a total area obtained by adding them up.

So far, we have been concerned only with maintaining a plant in a turgid condition and we have not mentioned growth or other changes which can occur. Now we must look more closely at the plant and attempt to understand its ways. The cell has already been mentioned, but we have not considered its contents in any way. If we place a cell under the microscope, we see that it is not just the bag of sugar we have so far been imagining. The more we magnify it, the more we see. The chlorophyll, for instance, is not spread evenly through the cells but is concentrated in tiny particles called chloroplasts—it is in these that the complex reactions of photosynthesis take place.

One region of special interest in the cell is called the nucleus. Usually this does not show any particular structure, but when a cell is dividing into two, a number of rather rod-like particles are seen if we treat the cell with a suitable stain. These are the chromosomes, and, as a general rule, they are constant in number and shape in all the cells of a given plant. The chromosomes contain almost all the information about the plant, so that it is possible, in theory at any rate, to grow a whole plant from any one of its living cells. Actually,

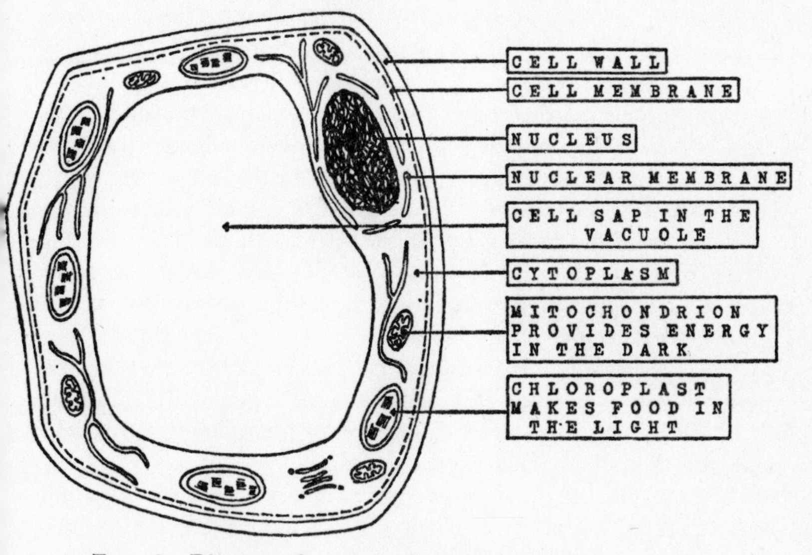

FIG. 16. Diagram of a plant cell highly magnified to show its contents

we find that only some cells are normally capable of growing in this way. Others become specialised so that they perform a particular job very well. The cells which divide easily are concentrated at the growing points of the plant and in various other regions, such as between the xylem and the phloem of woody plants where, as already mentioned, they form the cambium.

It may be wondered why cells, which contain the same information, can develop in different ways. The answer is, that the development of cells is controlled by certain substances which are produced, either within themselves, or by other cells. Botanists have known about some of these substances for a long time but others are only just being discovered. Some are known as plant hormones, and are, as we might expect, completely different from the hormones which are found in animals.

Actually, a large number of substances have been made by chemists which can act like the genuine plant hormones. Many of these have been developed as 'hormone' weedkillers, where they work by upsetting the normal patterns of growth.

Up to now, we have dealt only with the different parts of a plant as if they were totally separate structures but, in fact because of the production of hormones in a growing plant, they form one integrated organism. One of the simplest demonstrations of hormone action is that concerning apical dominance. When we look at many plants we find that they have a main stem which is growing strongly and only a few small side branches. In theory, a branch can arise in the angle just above every leaf (known as the leaf axil to botanists), but they do not do so because the tip (apex) of the strongly growing stem produces a hormone which stops the cells dividing. If we pinch out the tip, as we often do for some garden plants, the lateral (side) buds are set free from the hormonal control and start growing out to give a bushy plant.

The hormone which is responsible for this is known as 'auxin', and is involved in many other aspects of plant growth. The growing of roots towards a gravity stimulus and of the shoots towards light, are both apparently controlled by auxins.

It is not just the presence or absence of auxin which is important but rather the actual concentration. When a shoot is laid on its side, most auxin is found on the shady or lower side, and causes an increased rate of growth so that the tip grows towards the light or away from the ground. This response takes only a few hours to develop and once the growth has taken place it cannot, of course, be reversed. This is how spikes of bloom can be spoilt by the wind if they are not properly staked. They fall sideways, grow into an upright position again and then have a permanent 'crick in the neck'. Root growth is inhibited by high concentrations of auxin and therefore roots grow down instead of up under the same conditions.

At one time it was thought that auxin alone controlled the abscission (falling) of fruits and leaves. The theory for leaves was that when a leaf became old, its production of auxin was reduced, and this caused the subsequent development of a weakened zone of cells at the base of the leaf stalk, due to the reduced gradient of auxin

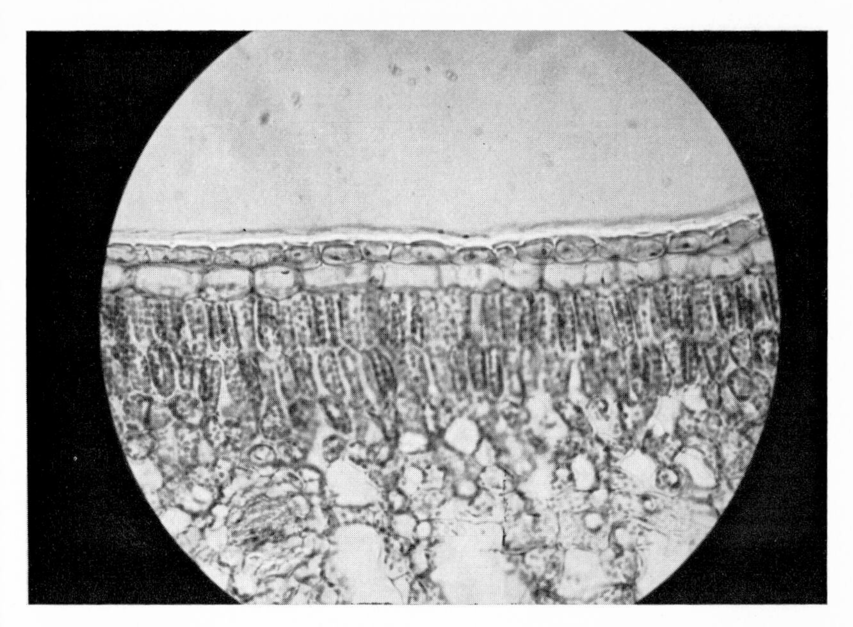

Plate 1. Vertical section of a green leaf of holly (*Ilex aquifolium*) showing the different types of cells (highly magnified).

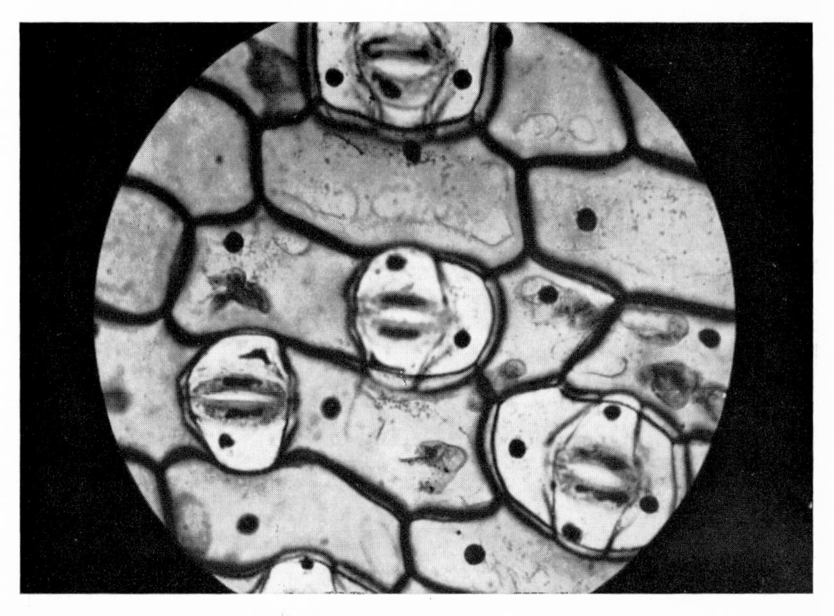

Plate 2. Highly magnified surface view of a leaf of *Tradescantia sp.* showing epidermal cells and stomata.

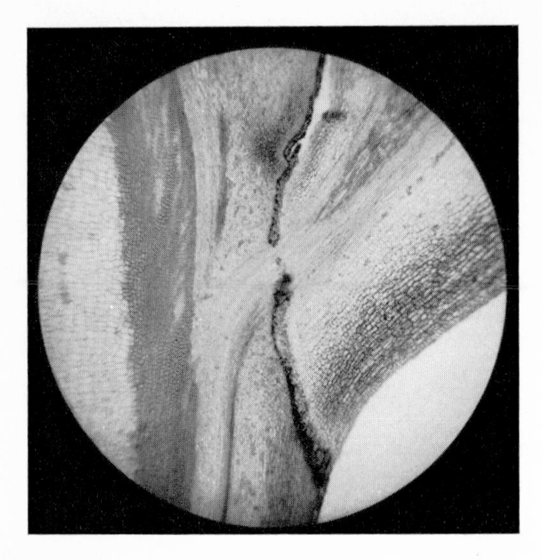

Plate 3. Longitudinal section of the junction of a stem and a leaf base showing the abscission zone prior to leaf fall.

(Photograph lent by Margaret McKendrick).

Plate 4. Nasturtiums which are laid on their sides soon develop bent petioles due to new growth stimulated by a response to gravity.

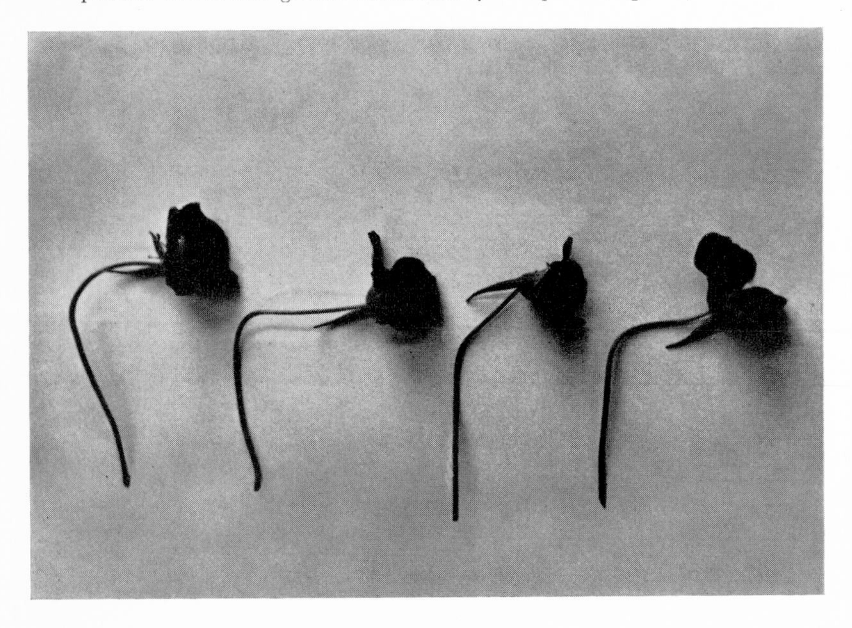

concentration from the leaf into the stem. Other experiments suggested that it was merely the total amount of auxin which was important rather than the concentration in the leaf relative to the stem. It is now known that there is also another plant hormone involved which has only recently been discovered. It is called abscisic acid, and is vitally concerned in abscission (see Plate 3).

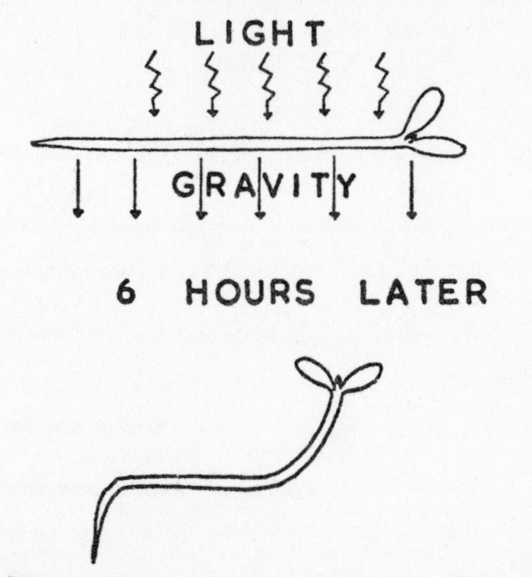

FIG. 17. When a plant is placed on its side the effects of gravity and light cause the shoot to bend upwards and the root downwards

Mention of old age brings us to yet another group of plant hormones called kinins. These are particularly involved in the division of cells and maintaining healthy growth. When a leaf gets old, it loses its green pigments and many of the nutrients which it contains. If kinins are put onto a leaf, the part where they are applied will stay green much longer, and the nutrients will not be lost. Unfortunately kinins are rather expensive to buy!

The gibberellins are another major group of plant hormones. One of their functions is to promote the elongation of stems and they are involved in the flowering of some plants. Certain biennials grow as a rosette in the first year and when they flower the stem elongates rapidly. A rise in the amount of gibberellin present is associated with this. A few plants can even be made to flower by the addition of gibberellin.

There remains to mention the, as yet, unidentified flowering hormones. These are still the subject of much research but have so far eluded exact identification. The search is complicated by the fact that conditions which favour flowering in one plant may prevent it in another. This is shown clearly by the response of plants to the length of the daylight hours. Some plants such as Border Carnations (*Dianthus sp.*) flower only in long summer days, while others will not flower in these and do so only in the shorter days of spring or autumn like the Florists' Chrysanthemums (*Chrysanthemum morifolium*).

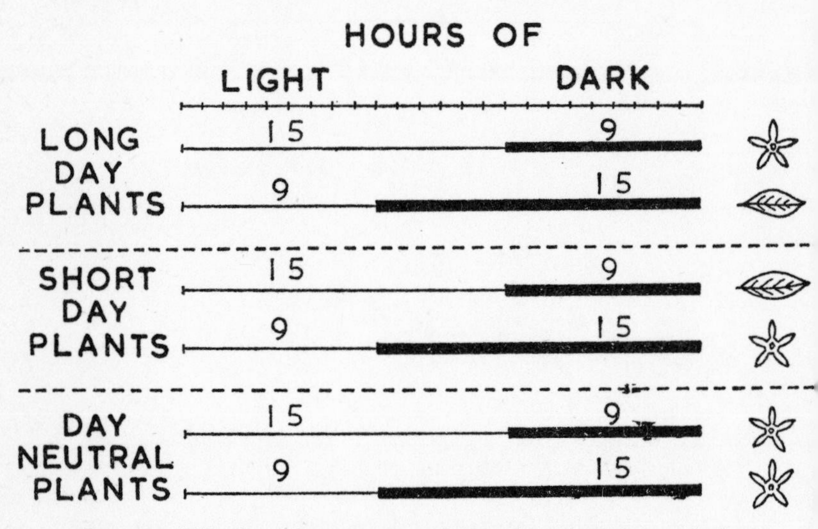

Fig. 18. The flowering response of long-day, short-day and day-neutral plants under different lighting conditions. A leaf indicates entirely vegetative growth

As far as botanists know the actual changes which take place in these different plants are the same. It is just that in some the long days switch on the flowering response while in others these same long days can stop it developing.

Of course the control of flowering is not as simple as is suggested by the above paragraph. Many other influences, such as age and temperature, can also control flowering in many plants. Certain commercially grown plants can be controlled so that they flower when required by adjusting temperature or daylength artificially. In this way the flower arranger can, for instance, have lily-of-the-valley (*Convallaria majalis*) and certain cultivars of pot chrysanthemums all the year round. Most plants are not so amenable to this form of treatment so that we still have seasonal changes in the flowers available to us.

Once the flower buds have been formed we cannot change them back to vegetative ones again. What we can do in some plants is to vary the size and stem length of the flowers by various cultural treatments. To grow strong and sturdy the plants need an adequate supply of nutrients. Naturally, if there are several buds on one stalk,

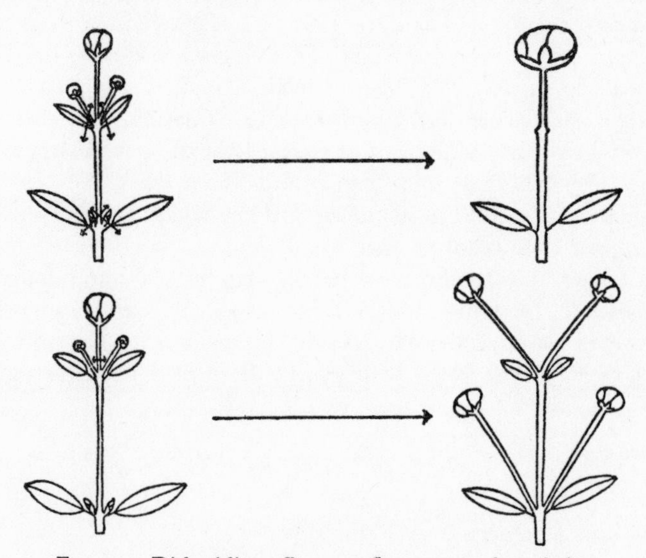

FIG. 19. Disbudding effects on flower growth and size

they will compete with one another and hence the familiar practice of disbudding. In some plants this means removing the lateral blooms and leaving only the strongest central one on each shoot. In other plants we may channel the strength into one or two shoots and pinch out all lateral branches to achieve the same ends. If the central flower bud is removed the lateral buds grow out to form shoots, each ending in a flower, and producing a 'spray' effect.

No account of flowers would be complete without mentioning the various opening and closing movements. In many cases these opening and closing movements are due to differential growth of the upper and lower sides of the individual parts of the flower. In some plants the flower never closes once it has opened while others show a distinct rhythm of opening during the day and closing at night. This may not always be due to the change in light intensity, but can sometimes be due to the lower temperature at night. Yet other plants may only open for one brief period before closing for ever.

A very striking closing of some flowers can sometimes arise for unexpected reasons. Perhaps the most common is due to minute traces of a gas called ethylene. This is present in small amounts in some older domestic cooking-gas supplies, but readers may be pleased to know that the underground 'natural' gas that is now often used contains no significant amount of ethylene. A more likely source of ethylene than leaks of domestic gas are various ripe fruits particularly apples. If a large number of apples are placed close together near susceptible flowers such as carnations, the result can be complete and permanent closure of the flowers within a few hours. This has been known to ruin a flower show. The ethylene gas acts like a plant hormone, but it is one which is volatile and so can spread from plant to plant.

So far we have been considering the growing plant and its general responses to the world around it. Now we must consider what happens when we upset the natural situation and cut the blooms we are going to use in our flower arrangement from the rest of the plant.

FURTHER READING

The Plant Kingdom by I. Tribe (Hamlyn).
Beginners Guide to Botany by C. L. Duddington (Pelham Books).

CHAPTER TWO

Cut Flowers and Leafy Shoots

It is a great shock for a plant when a part of it is cut and gathered for flower arrangements. But a shock which the remainder of the whole plant, left growing, can usually contend with because the sap rising to the cut surface dries in contact with the air and quickly seals up the wound. What was at first an open wound, and a possible place for the entry of bacterial and fungal diseases, soon becomes a healed scar. In some plants, a milky latex exudes and solidifies to a rubbery mass which covers the cut surface. The latex is contained in special elongated cells which are quite distinct from the xylem and phloem elements which transport the ordinary sap. If the cut across the stem is made cleanly, the healing is easier than in a torn and jagged place because the area of damaged stem is less. If the cut is made immediately above a bud, little dead wood is formed and an unsightly 'stick' is avoided.

The initial closing of the wound is due to the drying-out and death of the damaged cells, often accompanied by browning. The browning is partly a protective measure by the plant because the substances which give the brown colour also hinder the growth of moulds which might infect the cut. Subsequently the wound is closed by the growth of wound tissues. Botanists sometimes talk about a wound hormone which is released when cells are damaged and that is quite distinct from other hormones which may be involved later. The energy needed for this wound-healing growth comes from the food reserves of the plant and the healing may take several weeks or even years when a large branch is removed from a tree. The protective covering which grows over the wound is called a callus and is really just a mass of cells. When leaves fall off naturally from a stem, the process is rather different and will be discussed later.

The hormones produced by the growing tips of the shoots (which have now been removed with the cut shoots) have, until this time, concentrated on promoting the growth of the leaders (main stems) and acted as growth inhibitors to the side shoots, preventing their development. Removal of the site of production of these substances allows the lateral buds of the plant to sprout out. This is commonly seen when a hedge is trimmed and the side shoots grow quickly and 'green up' the hedge. In fact good hedging plants are chosen from shrubs which respond particularly well in this manner.

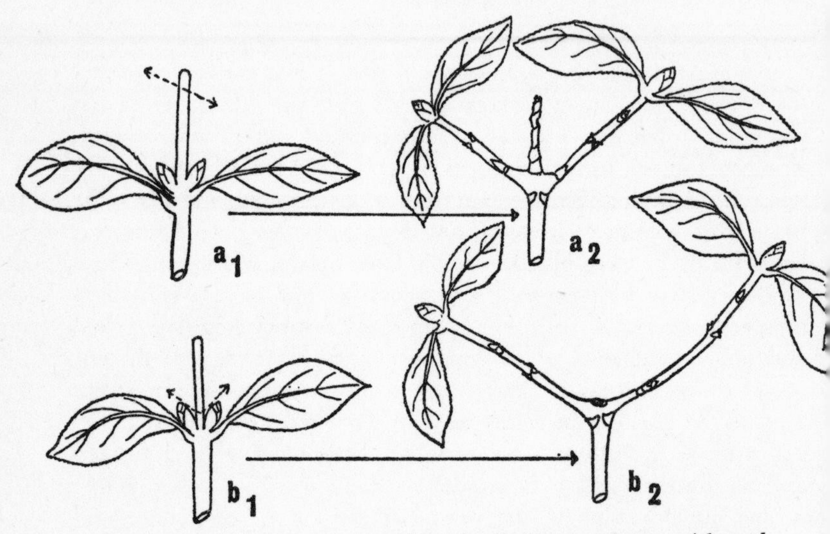

FIG. 20. To show the results of cutting hydrangea shoots. a_1) branch cut leaving several inches above a pair of buds growing in the axils of leaves; (a_2) shows the same shoot the following year. A dead stick is left between the new side shoots which terminate in a 'flower' bud. (b_1) A similar shoot showing the cut made immediately above the buds and leaves; (b_2) the same shoot shown the following year with no unsightly dead twig and strong lateral branches terminating in 'flower' buds

After the cutting the main plant will still be adequately supplied with mineral nutrients since its roots will remain in the soil. Provided that a reasonable amount of top growth is left, and this is usually the case, the plant will also be able to furnish itself with a supply of the products of photosynthesis and is usually not seriously harmed by cutting. It may even benefit by having its flowers removed before the great drainage of food materials occurs to supply the energy needed for fertilisation and the formation of fruits and seeds. Indeed, it is the practice of many careful gardeners to remove dying flowers immediately after their period of maximum attractiveness. This is known to decrease the strain on the parent plant and to promote further flowering.

Now, what of the severed shoot? Many of its essential processes will still go on but the necessary supply of nutrients and water via the roots has been stopped. Thus, the cut shoot will quickly die if something is not done to replace its normal supplies, particularly of water. Conduction of water, mineral salts in solution and manufactured foods takes place in the vascular system of plants. The tissues involved are initiated, developed and distributed in various very specific ways in different plants. Most of the plants in the world have been examined and the details of the structure and arrangement of their vascular and other tissues described. We can only mention a few simple features and further information on this vast subject—plant anatomy—may be obtained from the books listed at the end of this chapter. Of course to say that a plant has been described does not mean that we understand how it works. That is a much more difficult task.

Perennial herbaceous plants, those which die down to ground level at the end of each year, may be either sappy or woody, but they all display only one year's growth above ground just as do the annual plants which live for one year. Woody perennials add to their girth and to the terminal growth of their branches every year. building on the foundations laid in previous years. Thus it can be seen that the structures and needs of these plants will be different. Let us consider them in more detail.

The vascular tissues of herbaceous plants are grouped into bundles of conducting elements called vascular bundles or veins, whereas the conducting elements of shrubs and trees show larger masses of xylem and phloem elements which are added to each year. Ordinary 'wood' consists of xylem remaining from previous years—often even hundreds of years. The phloem of woody stems usually comes off as part of the bark when we pull off the outside covering of the woody stem.

If we look in detail at the structure of the xylem and phloem as revealed under the microscope we can see how perfectly these tissues are constructed to suit the performance of their functions.

The xylem consists mostly of elongated cells which form the channels along which the water flows. When these cells are quite distinct and simply lying end to end next to one another they are known as tracheids. In many plants the system is developed even more by the joining up of many cell 'units' which have lost the end walls so that they form a continuous system without any cross walls to hinder the flow of water. The tracheids and these vessels, although at first living cells with all the usual contents, soon die and lose all the cytoplasm and cell sap which was at first present. This makes them even more suitable to act as water channels. As will be remembered from Chapter One, the water in the transpiration stream is under tension caused by the evaporation of water from the leaves.

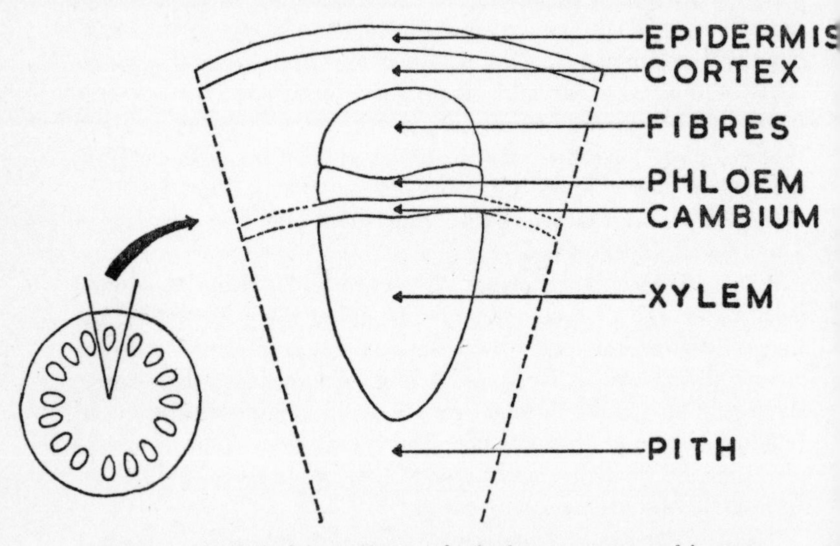

FIG. 21. A plan of a small part of a herbaceous stem and in transverse section to show the arrangement of tissues and a vascular bundle

The walls of the xylem elements are thickened in many different ways, presumably to resist the forces set up by this tension. Each type has its own name which is descriptive of the thickening—spiral, scalariform (ladder-like) and so on. In fact, the forces involved are such that if we accurately measure the circumference of a stem or tree we find that it is smaller when the transpiration stream is moving rapidly during the day. This daily variation in circumference is sometimes used to check that the vascular tissue of trees is functioning normally and not damaged by disease.

The elongation of the xylem elements is along the length of stems which is the main direction in which the water flows. Other cells in the xylem of woody stems are responsible for the movement outwards of water and dissolved substances to the outer parts of the stem. These are the medullary or vascular rays and they consist of cells which are somewhat elongated in the radial direction. Again the exact arrangement differs in different plants.

The phloem elements are also elongated but they have rather exceptional living contents which are, in a way which is not yet fully understood, involved in the movement of substances in the phloem. The individual phloem elements are interconnected end to end into continuous sieve tubes by the 'sieve plates' which have large holes like a sieve. Each element of the sieve tubes has a companion cell associated with it which seems vital to its proper functioning. Other cells in the phloem may have very thickened walls and provide strength to stems or leaves which have little xylem in them. These thick-walled cells are the phloem fibres.

In many plants thick-walled fibres occur in other degions of the stem or leaves and provide strength to the tissues. Most of the plants which dry out without losing their shape have these sclerenchyma fibres present in large amounts.

If we cut across a herbaceous dicot stem and examine the cut surface closely we can often see, with the naked eye, a ring of small spots which goes round the inside of the stem. These are the vascular bundles and they can be seen because they consist of the special cell types which have been mentioned above and these are quite distinct in appearance from the other cells which make up the stem. If we place the cut end of a shoot in water containing red ink, the ink will pass up the xylem elements with the transpiration stream and we can see the path of the vascular bundles even more clearly. The centre of herbaceous stems is often made up of soft pith, or may even consist of an air space so that the stem is hollow. The lining of the hollow central air cavity is much thinner than that around the outside of the stem and does not have the waterproof cuticle over it so that water can be unnaturally encouraged to enter some hollow stems from the inside, as mentioned in the next chapter.

In the stems of dicots which have become woody the separate vascular bundles, if they exist at all, soon join up sideways to form a complete cylinder of vascular tissue with a layer of cambium running through it to produce new xylem to the inside and phloem to the outside.

The 'stems' of monocots are often only associated with the pro-

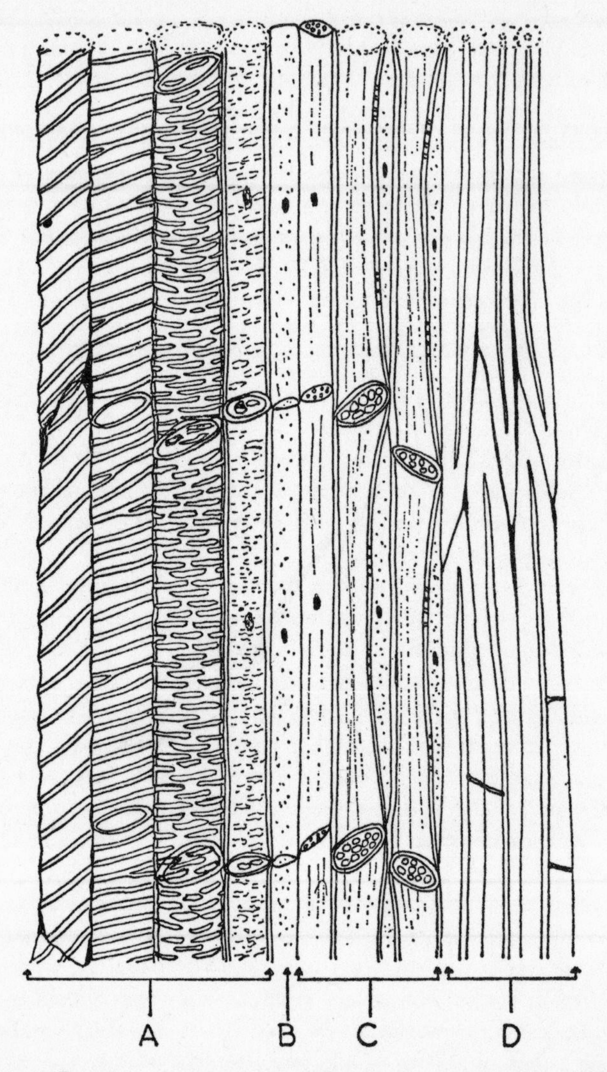

FIG. 22. Drawing of a highly magnified longitudinal section through a vascular bundle showing left to right (A) xylem elements, (B) the thin-walled cambium, (C) phloem elements and (D) pericyclic fibres

duction of the flowers and have a rather different structure. A typical stem, such as that of the sweet corn (*Zea mays*), has the vascular bundles scattered throughout the thickness of the stem and not organised into a cylindrical arrangement. A layer of cambium never develops so that monocots do not usually become woody and last from one season to another as stems above ground. The rigidity of the stems of grasses and other monocots comes from the presence not of xylem but of other tissues.

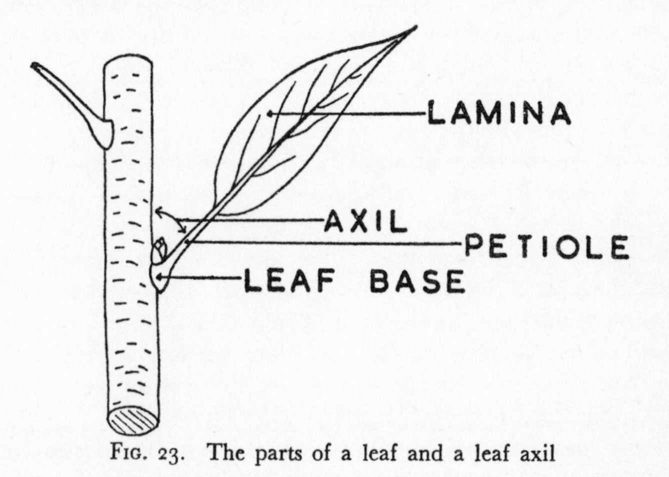

FIG. 23. The parts of a leaf and a leaf axil

The next part of the cut shoot which needs our attention is the leaf. Some plants, such as various biennial plants which form rosettes in the first year and many monocots, have their leaves all crowded together on extremely short stems. Such leaves which appear to arise direct from the ground are known as radical leaves in contrast to the cauline leaves which are obviously on stems.

The typical dicot leaf consists of two parts: the petiole (stalk) which runs into the leaf as the main vein or veins and the expanded lamina (leafblade). Petioles can usually be easily distinguished from stems by their shape in cross section, and the fact that a bud (perhaps very tiny), or a branch of some sort, arises in the axil of the leaf. The leaf axil is the region between the stem and the leaf on the side nearest to the apex (tip) of the stem.

The leaves of some plants have very long petioles, but in others, the leaf may be without any readily visible petiole and is then known as a sessile leaf. Occasionally the very base of the leaf stalk is obviously rather different from the main part of it and may be swollen up with rather large cells which, when turgid, cause the leaf to be held out straight but if these cells are flaccid, the leaf hangs down. Actually the leaves of some plants show a daily rhythm in the way in which they are held and at night are said to 'go to sleep'. The movements are related to the daily changes from light to dark but they will continue for a little while even if kept under constant conditions.

The leaves of monocotlyedons normally have the larger veins all parallel to one another in contrast to the usually reticulate (net-like) arrangement in dicot leaves. The leaf bases of grasses and some other monocots are greatly enlarged into sheaths surrounding the stems and often also the bases of the leaves nearer the apex.

Leaves join the stem at regions called nodes and the parts of the stem between the nodes are known as the internodes. A study of the nodes reveals a remarkable constancy in the way in which the leaves arise. The arrangement of the leaves and leaf shape provide useful diagnostic characters to help identify the species of a plant when no flowers are present on it. The botanical term 'phyllotaxis' is used when discussing the way in which the leaves are placed on the stem. If we look closely we can distinguish three basic types of leaf arrangement in different plants.

Firstly the leaves may arise in pairs opposite to one another giving the shoots a very regular appearance. If each pair arises vertically above the pair below, the arrangement is called distichous, but this is not very common. Usually the pairs alternate at right angles to those above and below, giving the decussate arrangement. This type of leaf arrangement is characteristic of almost all members of the family of plants known as the Labiatae to which *Salvia*, mint and many other garden and wild plants belong.

Rather less common are the plants which have a whorl of several leaves at each node.

Probably the most common arrangement is that in which the leaves arise one at each node. They do not, however, arise one above another but spiral around the stem from node to node. The spiral is most easily seen when looking closely at the tip of the shoot where the leaves are small and closer together. If this region is studied with a magnifying glass, it is often possible to see small lumps at the very tip which are the 'primordia' which will grow into leaves or flowers as the case may be. The shape of the flower primordia or a flowering apex is usually flatter than that of leaf primordia or vegetative apices, so that this gives a possible way of finding out whether or not the flowers have been initiated. This is of relevance when trying to force plants into early flowering. This is discussed in Chapter Four.

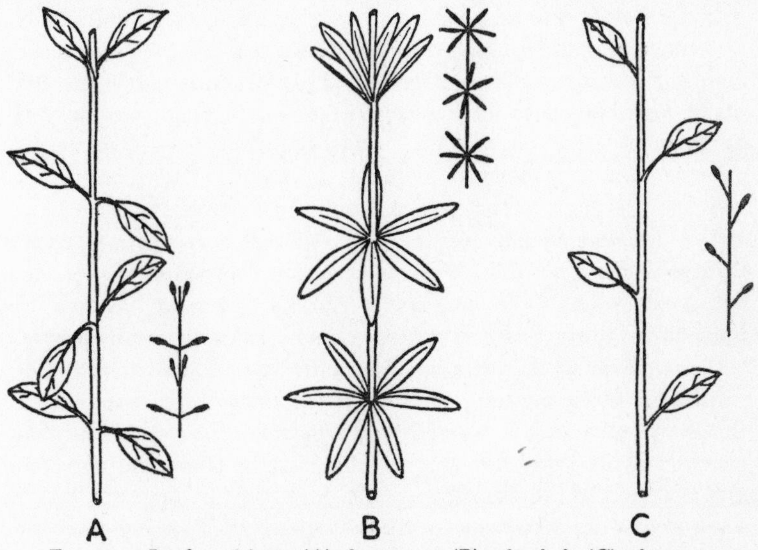

FIG. 24. Leaf positions. (A) decussate; (B) whorled; (C) alternate

Most of the plants which flower arrangers use have leaves. True, there are quite a number of flowers, mostly from bulbs and corms, which are borne on long, leafless stalks but even these plants usually have a rosette of leaves at the base of the plant. As was explained in Chapter One the leaves are the main organs in which the processes of photosynthesis go on. If leaves are absent, then other parts of the plants, usually the stems, are green and able to photosynthesise.

As we might expect the leaves reflect the conditions under which a plant grows. Plants which grow under moderate conditions which are neither too wet nor too dry are known as mesophytes which literally means middle plants. The leaves of such plants are the ordinary, flattened, non-glossy, leaves with which we are all familiar. This type of leaf loses a great deal of water and may wilt rather quickly. It is often preferable to remove some of the leaves from such plants. More water is then available in the shoot to keep the flowers turgid and in good condition. Less sensitive greenery from a different plant can be incorporated in the arrangement if needed.

Plants which grow in 'dry' places, or look as if they do, are known as xerophytes. We should remember that the term 'dry' means only that water is not readily available to the plant. This may be because there is little rainfall, or it may be because it is too cold for the plant to take in water easily. It may even be that plants which live near the sea have a rather similar appearance because the high salt content of the soil hinders the movement of water into the plant. These plants which live close to the sea or in salty regions are usually called halophytes. The leaves of plants from dry regions show xerophytic characters such as marked reduction in surface area, the development of a thick, waterproof, cuticle and the sinking of the stomata into the leaf tissue so that they can trap small pockets of moist air above them to slow down the water loss. Shoots with leaves which show some xerophytic characters, such as those of holly and ivy, are useful in arrangements because they last so long.

Some plants actually live in water and these are known as hydrophytes. These are not much used by flower arrangers, except perhaps for the flowers of the water lily. Perhaps we are all afraid that people will look for fish amongst the leaves! Some leaves which grow submerged are dissected up into dozens of tiny segments which allow the water to flow over them very easily. There is probably more scope for the use of some of these leaves in moribana designs than is actually made use of. Submerged leaves have very thin cuticles and hardly any strengthening tissues in them, so that when taken out of the water, they collapse easily and shrivel up very quickly. Floating leaves on the surface of water usually have special coatings to their upper surfaces to stop them being wetted. They also have a very well developed system of air spaces which helps to make them buoyant as well as allowing air to reach the submerged parts of the plant.

Thus far we have been concerned with describing shoots. What happens at the instant of cutting? As a stem is cut air rushes into the xylem tubes which normally carry water and dissolved minerals up from the root. Many bubbles of air form which increase in size every moment that the cut end is out of water. Even putting the cut end into water is not usually satisfactory in itself, because an air block has been formed which will not allow water to enter. Those who have turned the water off at the mains before leaving a house empty for a period and on returning, turned it on again only to find that an air lock prevents the normal flow of water, can envisage what has happened inside the plant. However much water is available to the plant, it cannot take the water in due to this air blockage. If the region of the stem containing the air lock is cut off under water, atmospheric pressure will cause water to enter and to connect up

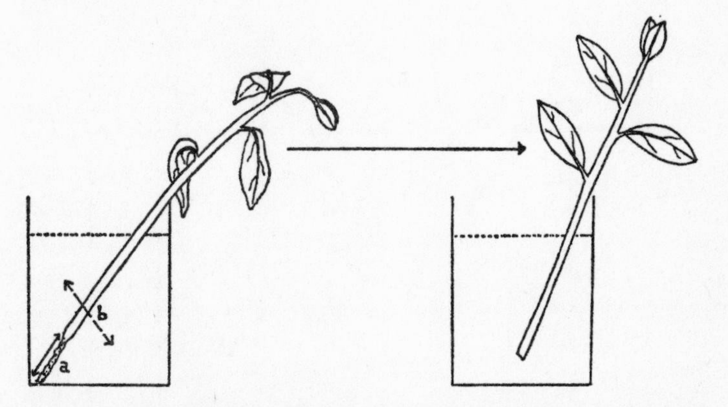

FIG. 25. Showing a stem flagging although the end is in water. (A) denotes the damaged base where an air block prevents the inflow of water to compensate for that lost in transpiration. (B) shows the position of a new cut made with a sharp knife under water. Soon the shoot will regain its freshness due to the unimpeded entry of water

with the normal sap within the xylem tissues and to flow freely through the plant. Another reason for cutting off the end again is to remove dead or dried regions which are blocking the water channels. This is particularly necessary if the cut end has been out of water for some time. The xylem elements are such thin, delicate tubes that they are easily damaged or closed up by blunt scissors so you should always use sharp implements. Other methods of dealing with cut stems are explained in the next chapter. Care at this stage is worthwhile because the life of the arrangement depends on it.

Many cut flowers and shoots respond to this second cutting under water. The flowers of anemones soon become fresh and fully expanded and last a long time if cut off under warm water. When anemones are bought as cut flowers, the base of the stems should be inspected for any area of brown flattened tissue which indicates that extensive damage has occurred to the conducting elements and that more than the normal two inches must be cut off. This, of course, considerably shortens the length of the flower stalks. Stems of tulips assume an upright condition and the floral parts become turgid and fresh after the warm water treatment, which is further helped if the tulips are stood in water up to their necks in a narrow milk-bottle-shaped container.

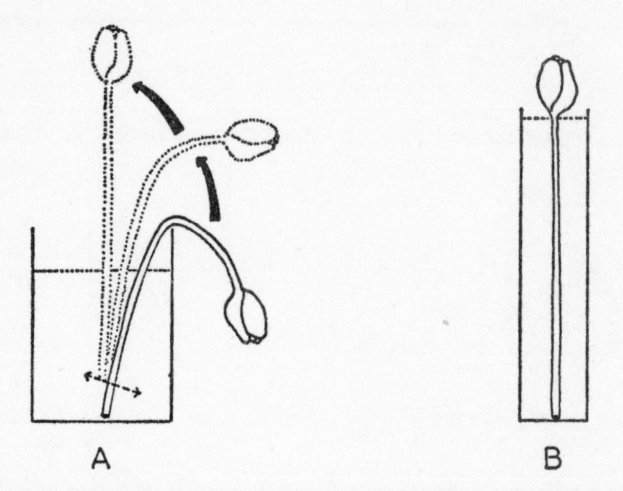

FIG. 26. Diagram to show rapid recovery of tulips after their stems have been recut under water. (B) standing up to their necks in water aids recovery, water being taken in along the whole length of the stem

All these comments reinforce the first principle of making sure that any cut shoots or flowers are given a satisfactory water supply as soon as possible.

The second major principle is to keep the loss of water from the flowers or shoots down to a minimum. If the movement of water (actually in the form of water vapour) out of the shoot is quicker than the movement of liquid water into the plant, then wilting will occur. This can only be prevented if the movement in is promoted and the movement out is hindered. This latter is best done by keeping the plant in conditions which favour the closure of the stomates (see Chapter One) and in conditions which do not favour evaporation. Thus darkness or dim light (causing the stomata to shut) and cool humid air (causing evaporation to be at its minimum) are ideal for keeping cut flowers fresh. Although most flowers should not be wetted, a fine spray of clear water used near cut shoots awaiting arrangement will promote crispness and long life. It is small wonder that some arrangements do not last long in the hot dry air of our centrally heated rooms.

A word about relative humidity is perhaps not out of place here. As we all know, water evaporates into the air, and if we have an open dish of water in a closed container, the water will go on evaporating until the air is saturated and can hold no more water. This is the state of 100% relative humidity, and if less than this amount of water is present there will be the corresponding lower percentages of relative humidity. The actual amount of water held at 100% relative humidity is not fixed however because if air is heated, it can hold more water so that what was fully saturated air at fifteen degrees centigrade will not be fully saturated at twenty degrees centigrade unless more water is evaporated into it. Conversely, if air is cooled its relative humidity will increase and may even reach the point at which drops of liquid water start to form. Since a plant is almost ninety per cent water, it gives up water to the atmosphere quite readily. The lower the relative humidity of the air, the quicker the plant will dry out.

So, flagging can be warded off by carefully understanding the problems of the cut flower, and providing better living standards. Flower shops are cool, shady places and cut flowers are kept overnight in deep water containers in cool dark places. Although these facts are recognised by most people, they are often taken for granted and their real significance not fully understood.

With the question of water cleared up, the next problem for a cut shoot is the supply of nutrients and food. When attached to the rest of the plant, minerals and other substances pass up from the roots dissolved in the water. If we wanted to grow a plant in water rather than soil, we should have to do two things. First, we should have to aerate the water because without oxygen the roots do not readily take in water and nutrients. Secondly, we should have to supply the mineral nutrients which are normally present in water which is mixed with soil. It is possible to do both of these fairly easily. In fact, some growers produce crops of tomatoes and other plants without soil. The mineral nutrients must be present in very controlled amounts however, otherwise the plants do not develop properly.

Fortunately, the flower arranger does not really have to bother about either of these two problems. Since no roots are present the supply of oxygen is not critical. The water moves into the cut end because it is sucked in by the tension caused by evaporation from the leaves into the air. The cut flowers and shoots do not normally last long enough to become short of mineral nutrients. There will not be much new growth after cutting, and what little there is can be supported by nutrients drawn out of older leaves or other parts of the shoot. The roots also produce complex substances many of which have not been fully identified yet by botanists. It is the lack of these which is probably largely responsible for the death of the cut shoot if we have attended to all its other needs.

As discussed in Chapter One plants manufacture their own food by photosynthesis. The amount of food in a cut shoot is likely to be greatest in the early evening when there has been little time for it to have moved out from the green leaves into the roots and other parts of the plant. The amount of food in the shoot may be an important factor in keeping it alive. In the dim light indoors, little photosynthesis can take place to produce new food. In any case we may have removed the leaves and some flowering stems may not have had any to start with. To put the shoots in conditions which would favour photosynthesis, such as under bright lights, would be both difficult to achieve and also not entirely satisfactory. The opening of the stomates, allowing gaseous exchange, would encourage water to be lost and any heat from the lamps (which could be considerable) would further aggravate the problem by lowering the relative humidity around the plant. Loss of water would then cause the stomata to close and hamper photosynthesis.

The obvious answer would be to provide sugar for the shoots to use in the water in which we stand them. The problem is firstly that sugar naturally moves mostly in the phloem rather than the xylem elements. The taking-up of water by the shoots is almost entirely as the transpiration stream moving along the xylem tracheids or vessels. Provided there is not too much sugar in the water we can get some into the plant via the xylem. This, in any case, is the route which plants use to transport dilute sugar solutions in the spring when the sap is rising. In this way some of the food reserves stored in the roots are sent up to the expanding buds.

The problem with placing sugar in the water is also that its presence encourages the growth of micro-organisms. If they just lived on the excess sugar this would not matter, but unfortunately many of them produce slimy substances which can block up the ends of the xylem elements and upset the water supply into the cut shoot. Furthermore some of the micro-organisms may be encouraged to attack the cut end of the shoot and rot the base so that again water uptake is upset. Carried up in the water the microbes could also spread to every part of the shoot and cause death and decay to set in all over it. Yet again some micro-organisms can produce chemicals which are toxic to the shoot even if the microbe itself does not actually rot the tissues of the shoot. The ways round this are briefly mentioned in the section of Chapter Four, which deals with additives to the water.

Even without added sugars flower water often quickly smells bad. This is due to the natural processes of decay promoted by bacteria in the water working on the cells of the plant and the sugars which have leaked out. They produce certain gases which smell objectionable. Bacteria work on soft plant material very quickly and therefore leaves should be removed from the lower submerged parts of the stems. Simple ways to cut down the flow of sugars from the cut stems are mentioned in Chapter Three. As bacterial growth and action are controlled by temperature and a rise of a few degrees can speed up the reactions, it will be appreciated that the warm conditions of a living-room may be ideal for bacteria even if not for the flower arrangement!

It is worth remembering that the food reserves of the shoots and flowers are used up more rapidly at higher temperatures so this is an additional reason for keeping cut flowers and shoots in a cool place so that they can conserve their food and hence their energy.

Try as we will, every cut shoot will eventually die unless it produces roots and grows into a complete plant. That, of course, is the way in which we propagate plants by cuttings. There is one other stage which we can recognise in the life cycle of many plants. This is the abscission (or falling-off) of the leaves, flowers and fruits which occurs naturally.

The exact causes of leaf and petal fall are by no means fully understood. Petal drop is known to be intimately connected with the shedding of pollen from the stamens. Once the pollen has been shed one of the main functions of the petals, which is to attract pollinating insects, is over. Another function of the petals is to protect the stamens. It has been found that if the stamens are removed from the flower when they are quite young, or if they are 'fixed' with hair spray to prevent them from splitting and shedding their pollen, or if, as happens in some cases, the flower has no stamens, the petals tend to stay in position longer. Some of these practices are too laborious for general use but have been found to be effective when tried experimentally.

Double flowering cherries, the double *Kerria japonica* and similar flowers retain their petals much longer than the single flowered forms of these plants. Double flowers, of course, have many more than the normal number of petals. This may be due to various causes. In some cases the stamens seem to become converted into petals, whilst in others it seems that the number of petal primordia increases. Other double flowers such as those of stock have both stamens and pistils missing. The result of these various possible changes is that double forms are rarely as fertile as single forms of the same species. Many are completely sterile. From the flower arrangers viewpoint this means that many of the normal changes associated with the shedding of pollen, pollination and fertilisation do not take place. The double flowers therefore last longer. In the largely sterile wisterias where pods of seeds are rarely formed, the whole flower falls at maturity, whereas in the related laburnums, which set seed readily, the petals fall separately leaving the pods to develop. Thus petal fall here seems to be related to seed set.

Leaf fall often takes place at certain times of the year. The shedding of the leaves of deciduous trees in autumn is a familiar sight. Thus in some plants, the leaves live for a much shorter time than the stems or roots. When the leaves are shed, they are usually not completely dead but have become senescent—which is another way of saying that they show the signs of old age. Even the leaves of 'evergreen' trees do not live as long as the tree. It is simply that they do not all fall off at once as do those of deciduous trees.

The separation of leaves and some petals is preceded by the formation of a special abscission layer and other changes at the region of the leaf or petal base.

If a branch of a deciduous tree is killed suddenly, the leaves on that branch will not fall off in the autumn because these special changes have not occurred. The material holding the cells together in the abscission layer softens and forms a jelly-like substance. The leaf or petal is then left attached only by the delicate veins which easily break if shaken by the wind or the hand. In many cases a protective layer has already been formed on the inside of the abscission layer so that when the leaf or petal falls, an almost perfectly healed wound is left. Only the vein ends remain to be sealed up to give complete protection.

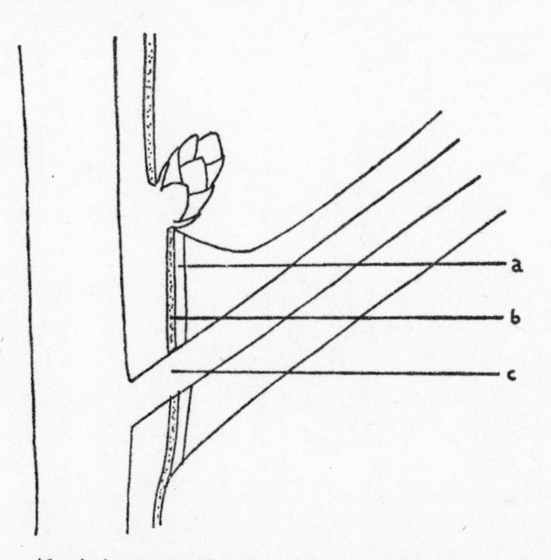

FIG. 27. Abscission zone showing (a) the region of the separating coils and (b) the corky tissue which forms the scars on the stem side. The vascular bundles (c) is not sealed off

Leaf scars such as these are the familiar horse-shoe-shaped structures seen on the winter twigs of horse chestnuts where the 'nail-holes' mark the position of the vascular bundles. On other trees the shape of the scar similarly reflects the shape of the leaf base.

It is not known why poppy sepals fall as the flower bud opens, why apple petals fall and the stamens and sepals are retained on the fruit, why rhododendron flowers stay attached to the plant (and should be removed by hand), why shrivelled flowers of some *Erica* species remain attached to the plant for many months or why the leaf blades of Virginia creeper are shed first and then the stalks are shed later. There is room for much research into these interesting problems.

There does seem to be a certain relationship between the life span of flowers and the number which are produced. The individual flowers of large inflorescences usually never last as long as does the flower on a plant which produces only one. Orchids are usually quoted as amongst the longest-lived flowers some lasting over four weeks provided that pollination does not take place. If they are pollinated fading of the flowers and other changes rapidly occur.

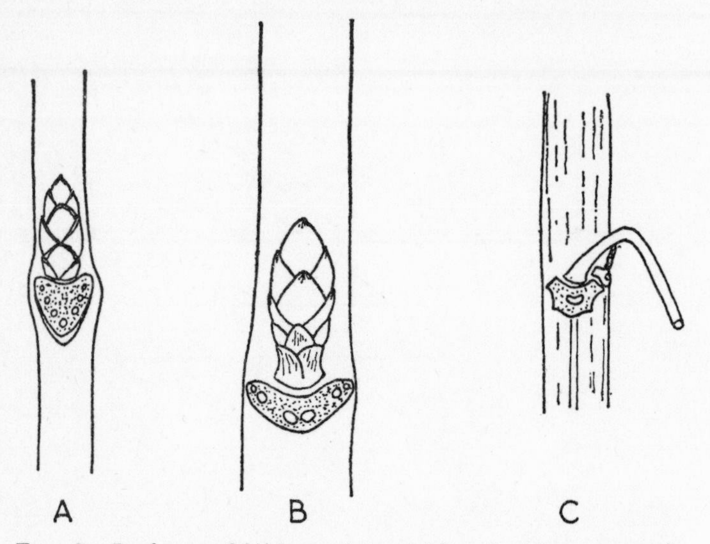

FIG. 28. Leaf scars of (A) horse chestnut (*Aesculus hippocastanum*), (B) sycamore (*Acer pseudoplatanus*) and (C) Chinese lantern (*Physalis franchettii*)

We do know however that hormones are involved in leaf and petal fall. When pollen is transferred to the receptive stigma of a flower, a tube grows out from it and fertilisation occurs. After fertilisation there is an increase in the amount of auxin and this causes the fruit to grow. At the same time, however, the remains of the flower may fall off and naturally this is of considerable concern to flower arrangers. The response is very complex since before fertilisation these other floral parts themselves produce auxin. Removal of the female parts of the flower (the ovary) before fertilisation normally causes the flower to drop off and not stay on longer as might be thought. Application of auxin to flowers however, often helps to keep them on the plant. We are only just beginning to understand the role of the substance known as abscisic acid which is known to be produced in senescent leaves and to stimulate abscission.

Cut shoots of flowers sometimes open very well in water if the conditions in which they are arranged are approximately the natural growing conditions and although it is not desirable for our arrangements to 'grow' in vases it is good to see buds opening as we remove dead or dying flowers farther down the stem. The forcing of buds and flowers 'out of season' is discussed in Chapter Four.

Whatever we do or say, all the plants in a flower arrangement will eventually die. Usually we discard them long before that happens, but it is worth remembering that there may be one or two which will dry well if taken out of the water. We must not anticipate Chapter Five too soon. For the moment let us try to keep our plants alive!

FURTHER READING

Life of the Green Plant by A. W. Galston (Prentice-Hall).
The Living Plant by P. M. Ray (Holt, Rinehart and Winston).
Evolution in Plant Design by C. L. Duddington (Faber and Faber).

CHAPTER THREE

Some Simple Hints and Possible Botanical Explanations

It should be remembered that although by the correct treatment it is possible to revive plants that have wilted it is not possible to bring plants back from the dead! It is therefore worth mentioning some of the differences between these two conditions.

Wilting is a sign that a plant has lost more water than it is taking in. Because the cells are not fully turgid, the leaves have a soft, flabby look and feel. If there is not much lignified tissue the leaves and even the stems may flop over. They will still remain green and will show only the slightest hint of turning a darker colour. If treated in sufficient time the plants will recover in a matter of a few hours.

If wilting is allowed to continue for too long a time, delicate leaves will start to show signs of death. Shrivelling of the tips and edges of the leaves is one. A marked darkening is another. Finally the leaves will become very dry and brittle.

Plants, just like humans, can be acclimatised to various conditions. This means that, quite apart from differences between species and cultivars, even the same plant grown under different conditions may be more resistant to water loss. If it has been brought up in a cool shady spot, it will be less resistant to desiccation after picking than if it had been grown in a dry sunny position. Even the two sides of a single shrub may show differences. The leaves from the sunny side may be able to lose more water before showing signs of distress. Certainly it is sometimes easy to see a clear difference in structure between 'sun' and 'shade' leaves. This does not mean that plants will benefit from being short of water for a couple of weeks before picking! What is meant is the long-term growing site of the plant. If you have a special display in a few weeks' time and the weather is very dry, you should water the plants adequately to ensure that they are in a fit state for picking when needed.

Quite apart from these cultural differences, it should be remembered that some flowers do not last long when cut, whatever is done to them. Old-fashioned roses, poppies, lilac and mimosa are well known for their beauty when growing and their disappointing appearance in a cut-flower arrangement. This is due to a combination of anatomical and physiological features which have not been altered by special breeding for the flower market. The really good cut-flower cultivars used in the trade have been selected because they last well and stand up to travelling. They are grown specifically with the object of producing cut flowers and are clearly a rather different proposition from a few blooms taken from a plant which is primarily intended to decorate the garden.

CONCERNING THE COLLECTION OF FLOWERS, STEMS AND LEAVES FOR USE IN DECORATION

1. It is usually considered best to gather flowers at the end of the day.

Explanation:

Plants manufacture their foods during the daylight hours using energy from the sun. These food materials are often distributed to other parts of the plant or to the storage organs such as swollen roots, during the hours of darkness. Early evening is therefore a good time to collect floral shoots containing the maximum food reserves, thus enabling these detached parts to survive longer.

It is also known that cells rich in dissolved substances resist desiccation better than those with weaker concentrations.

Simply from the water content viewpoint it is better to collect during the night because the water content of the shoot is highest. During the hottest parts of the day, water is lost by the shoots more rapidly than the roots can take it in. This may even cause temporary wilting. At night this deficit is made up. As flower arrangers we usually provide water but not food although this can be done in the special circumstances which are mentioned in the next chapter.

There are, of course, some flowers such as roses which appear to be best gathered at midday. The reasons for this are not clearly understood, unless the reason is merely to prevent one's fingers being pricked in the half-light!

2. Flowers should be cut and collected into a flat basket or trug to avoid holding the bunches.

Explanation:

Obviously physical damage to flowers must be prevented. Loose piling in a flat trug keeps some still, moist air around the shoots and so keeps down the loss of water by transpiration. This would be considerably greater if the shoots were carried around clutched in the hand and waving about in the air. Once cut the shoots should be kept out of the sun and taken into a cool spot as soon as possible for the same reasons.

3. The stems of cut flowers should be placed in water as soon as possible.

Explanation:

As explained in Chapter Two, the water in the stem is under tension when it is attached to the growing plant and as soon as it is cut off air rushes in. Preferably the stems should be cut under water but this is usually impracticable, so the next best thing is a bucket of water close at hand. Large bunches of flowers should be collected a few at a time, placed in water and then a few more gathered.

For reasons again not understood, peonies are one of the few flowers that are said to benefit from being left out of water for a few hours immediately after cutting.

4. Acanthus is best gathered when the topmost flower (which is small and not of the same artistic value as the attractive bracts) is just opening. The same is probably true for many spire shaped inflorescences.

Explanation:

At this stage the hormones causing the lower parts to die off have not been released and the spike is in its prime. Incidentally after using as a cut flower the whole spike of Acanthus can be dried off and used for years as the bracts dry slowly losing very little of their shape and colour.

5. Gladiolus spikes last best if the top three buds are removed.

Explanation:

Once all the buds have opened the lower flowers die off quickly. Hormones are released causing senescence. Also, the small, young flowers would drain food from other flowers to supply the energy for their own opening.

6. The lowermost flowers in spikes should be removed as they die.
Explanation:

If the flowers have been pollinated or normally develop without this they will start to drain the food reserves of the cut shoot to assist in the formation of the seeds and fruit. This will be even more harmful than if we allowed it to happen on the whole growing plant.

7. The seed heads of clematis should be picked just before they are fully ripe.
Explanation:

Picking at this stage prevents the final ripening from taking place and the consequent releasing (abscission) of the feather fruits. If the stems are further placed in glycerine and water solution (which is described in detail in Chapter Five) the glycerine enters their cells and prevents the final ripening. Other fruits also sometimes benefit from early picking provided they are rigid enough not to shrivel up.

8. Many flowers benefit from being picked just before they are fully mature.
Explanation:

This gives extra time before the arrangement will be at its peak. The main problem lies in the ability of the flowers to continue their opening when they are placed in water. If picked when mature the shock of being cut, which usually involves temporary wilting, may be sufficient to cause the death of the flower although it might have lasted for several days on the plant.

9. Sagging forced tulip stems may be rolled in newspaper and stood in warm water (comfortable to the hand) for 1–2 hours before arranging.
Explanation:

This gives support and damp air around the stems which are able to take in moisture and regain their turgidity. The roll of newspaper prevents the xylem vessels from being pinched off by the hanging down of the heavy flowerhead. Some flowers with soft stems may even benefit from a thin piece of floral wire pushed up the centre of the stem to the base of the flowers. This has the same effect of keeping the main water channels open.

CONCERNING IMMEDIATE TREATMENT OF THE CUT FLOWER.

1. Long spikes of flowers or flowers with long, soft stems should not be kept lying down overnight or even for several hours (see Plate 4).
Explanation:

Many flowers, particularly those with tall, upright spikes, respond to the force of gravity by growing away from the centre of the earth. If the plant is knocked down or the cut flowers laid on their sides they will have started to respond after a few hours or even less and since the curvature of the stem is caused by growth it cannot be reversed. The best that can be done is to reverse the position of the flower with respect to the earth so that the second response can partly cancel out the first. It will not do so completely however so that a wiggle in the stem will be inevitable.

The plants which show this type of response most strongly are those which have straight, upright spikes of flowers although even single flowers such as the humble nasturtium may show it.

2. The cut ends of many herbaceous and woody stems should be immersed in boiling water whilst a slow count up to thirty is made. The stems should then be left in the hot water, as it cools, for half an hour or even more. The foliage and flowers must at all times be protected from the steam by wrapping them in a damp cloth or paper.
Explanation:

Boiling water kills the cells which are immersed in it, and prevents the cells growing into a callus which would seal off the end of the stem. Browning and other detrimental changes in these cells are also prevented. By coagulating the proteins in the cells it lessens the living barrier to water movement. Also, because the phloem cells are killed, it prevents them from chanelling out sugars and other nutrients from the stem into the water around the stems. These solutions are known to promote the growth of micro-organisms in the water. Since the boiling water also sterilises the cut end of the stem, the actual number of micro-organisms which are there to start with is also reduced. These micro-organisms produce slime which tends to block the water channels in the stem and also cause the water to go off more quickly.

The reason for leaving in the hot water is partly to ensure a thorough job and also to promote the flow of water because warm water moves more easily than very cold water. This gives the cut flower a good start in making up any water shortage.

3. About two inches should be cut cleanly off the stems while the end is under water. This is particularly so for flowers which have not been picked within the previous few minutes.

Explanation:

This removes damaged and dried impermeable layers which have formed at the end of the original cut. It also helps to remove any air block which has been set up by the inrush of air into the xylem when it was first cut. This inrush, of course, compensates for the difference in pressure inside and outside the plant in its natural state where the water in the xylem is under considerable tension. Leaving plants around out of water will cause the air blocks (or bubbles) in the xylem to grow larger as more water evaporates from the leaves and flowers.

4. Rose stems and some other woody stems benefit from having the extreme tip of the cut stem held in a flame for a minute or so. The rest of the stem and flowers, must, of course, be shielded from the heat. The flowers will then last longer.

Explanation:

This is a similar tip to the use of boiling water but is more drastic. The heat kills the cells and prevents them growing into a protective callus or otherwise sealing the cut end. The large area of dead cells allows water to enter the base of the stem more readily. Killing the cells also slows down the leakage of food into the water of the vase.

5. Stems which exude a white, milky liquid when cut, such as euphorbias and poppies, benefit by having the cut end of the stem burnt as in the above tip.

Explanation:

Apart from the above explanation there are further points of importance with these plants.

The white, milky liquid is called latex and is a solution of rubber which will form a water impermeable layer when it dries, completely stopping the inflow of water. The latex often flows from long thin cells which form an interconnected network throughout the plant. If we merely cut off the dried end the latex would flow out again and we would still have the original problem. By killing the cells at the cut surface we prevent any further flow and the heat chars the rubbery material already exuded so that it is no longer impermeable to water.

6. The end two inches of woody stems are often hammered before being placed in water.
Explanation:
The hammering damage greatly increases the surface area which is available for the entry of water into the stem. The area exposed is so great that it will be some time before it is all covered over with an impermeable layer. Although there will be an initially greater loss of plant food from such a hammered stem, coming from the damaged cells, the area of living phloem exposed will not be much greater than if the stem had been cut across.

7. The cut ends of flower stems are sometimes plunged into melted wax. This seems to prolong the life of the flowers.
Explanation:
The wax completely seals up the ends, preventing sugars from leaking into the water where they would promote the growth of microorganisms. It will also lessen callus formation and other changes at the cut end. It is probable that the water can enter the stems above the waxed regions, possibly through the epidermis or, if it is a woody stem, as is usual with this treatment, then through the lenticels which form holes in the waterproof cork layer.

8. Anemones and many other plants need warm water and droop hopelessly in cold water. They also often fail to open, which is disastrous with flowers which are usually bought in the bud stage.
Explanation:
Most living and even non-living processes take place better and faster at warmer temperatures. Obviously there is an upper limit for those which involve living things because the delicate enzymes which help with the reactions are destroyed by excessive heat. In this case we are talking about a temperature which is comfortably warm to the hands. A rise in temperature makes water more 'runny' (in scientific terms it reduces the viscosity of water—just as heating treacle makes it more runny although the difference with water is not so obviously). This means that the water can enter the cut end

of the shoot more easily and spread up the plant and into the leaves more quickly.

9. Peonies respond best if left out of water for as much as six hours before arranging. During this time they should be laid in a cool place. Some people cut them at night and collect the next morning from the garden or fields.
Explanation:
This is a treatment which goes completely against most accepted ideas. It can only be assumed that the peony has something very unique about the way it grows and moves water through its stems and shoots.

10. Poinsettias are often arranged not as cut flowers but on their own roots, which are immersed in the water. The same treatment is sometimes recommended for various pot plants in which the soil can be easily washed out of the roots.
Explanation:
Although the flower arranger normally uses shoots and flowers without the roots, we should remember that under natural conditions the roots form a vital part of the plant. Not only do they supply water taken in from the soil to the shoots, but also they produce various hormones and other substances which are beneficial to the shoots. We normally only remove them for reasons of convenience and because roots immersed in non-aerated water do not work properly. If the roots can be packed in something which lets air get to them, such as damp spagnum moss, they will remain quite healthy for some time and the plant can be replanted when the arrangement is finished. Such shock treatment should not be repeated too often in one season.

11. Lupin spikes with hollow stems last well if upended, filled with water and then plugged with cotton wool and then immediately turned the right way up again.
Explanation:
This, besides keeping the whole plant fresh prevents the premature falling of the flowers. The cotton wool acts like a wick drawing more water into the centre of the stem from the water in the container. The water probably enters the vascular tissue of the stem more easily from the rather unprotected inside of the hollow stem than from the protected outside.

12. Many flowers regain freshness if stood up to their necks in cool or slightly warm water for several hours prior to arranging.
Explanation:
Many plants are known to take in water by the epidermal cells which surround the stem and leaves. In fact many plants will even take in nutrients in this way and this is made use of by gardeners when they use foliar feeding of special very dilute fertiliser solutions. This intake of water supplements that coming from the cut base of the stems. The support given by the water to delicate, non-woody stems which have little strength except that due to their turgor, will help to prevent closure of the vessels due to mechanical collapse of the stems. Warm water helps for the reasons already mentioned in 8 above. Many plants with delicate leaves must not have them immersed in water because they will quickly die and start to decay.

13. If the leaves are removed from flowering stems the flowers last much longer. (Foliage may be used separately from the flowering stems).
Explanation:
Defoliation cuts down the transpiring surface and less water is lost by the shoot so that more is available to the flowers to ensure their turgidity. Of course any leaves which would be submerged are normally removed anyway because of the danger of decay. It should also be remembered that some plants which have good flowers bear leaves which just do not stand up to being cut and placed in a vase. In these cases the foliage is replaced by that from another type of plant.

14. Very young leafy shoots wilt quickly after cutting and make poor arrangements.
Explanation:
This is always a great pity as the young leaves of oak, acers and many other shrubs and trees are lovely in their natural state. The young parts however, are poorly lignified and rely on the turgor pressure caused by the water to maintain their shape. The really high turgor needed is quickly lost if the rate of transpiration exceeds the rate of water intake, which of course, it often does. In addition, the protective epidermis of young leaves is less developed than when they are more mature. Young leaves are also partly dependent upon

Plate 5. Display of some grasses. This photograph, supplied by Mr M. T. Fessey, is just a small part of Dr Hubbard's exhibit of ornamental grasses at a 1970 R.H.S. Show which won him a Gold Medal of the Society. There are hundreds of different kinds of grasses which can be used in flower arrangements.

Plate 6. A slatted wooden plant press with adjustable straps similar to those used by botanists.

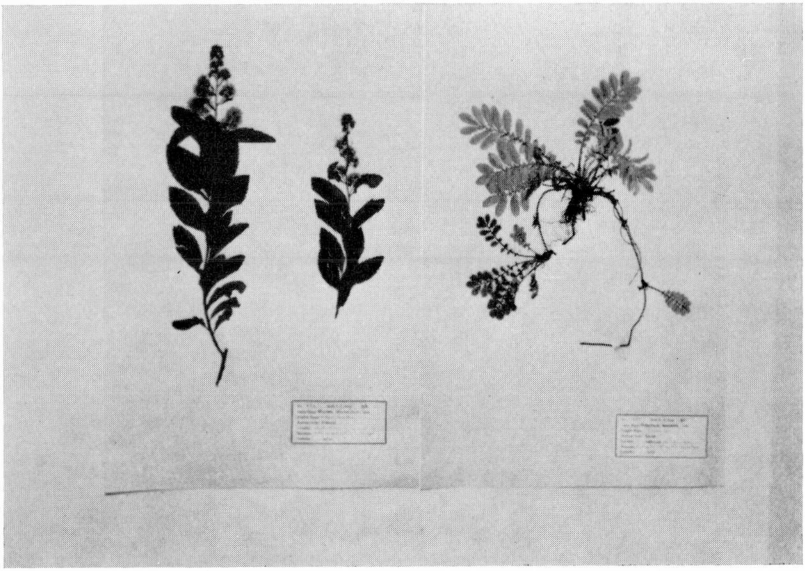

Plate 7. Typical herbarium sheets. Note that the plants are carefully spread out. The label gives information about the name, date and place of collection and other relevant details.

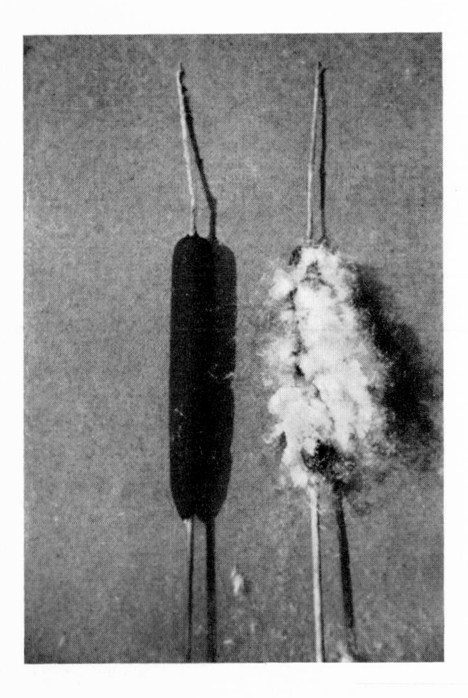

Plate 8. The compact, mature head of the bullrush lasts for some time but eventually explodes to set free the feathery fruits.

the food produced by the older, more active leaves or brought up from the reserves in the stems or roots.

15. Before arranging violets, submerge the whole flower in cool water for an hour. This results in fresher, long-lasting blooms.
Explanation:

It appears that the outer tissues are able to take in water through a cuticle which is less restricting than that of many other plants. Violets grow naturally in shady, moist places and the response to over-all dampness is probably the result of this. They also have a fairly large surface area compared to the thickness of the stem which makes the taking in of water by the normal route of the cut stem end hardly sufficient for their needs. This technique of total immersion is quite unsuitable for most flowers which are completely ruined by total submergence. One of the reasons for damage under water is the lack of oxygen and those flower arrangers who have experimented with the glass globes which can be completely filled with water, will remember that the use of hydrogen peroxide in the water is a vital step. The hydrogen peroxide provides a certain amount of extra oxygen in the globe as well as greatly hindering the growth of micro-organisms which would cause decay of the submerged flowers.

CONCERNING SPECIAL TREATMENTS TO PROMOTE THE LONGER LIFE OR BETTER APPEARANCE IN FINISHED ARRANGEMENTS

1. Damp tissue paper placed over hydrangea heads brings wilting or dried-up 'flowers' (which are actually petal-like bracts) back to freshness.
Explanation:

By increasing the relative humidity outside the bracts the tissues inside them can take up water all over the surface instead of losing it. A reasonably high relative humidity around the shoots is important in obtaining a long life for any arrangement. Warm, drying draughts are to be avoided at all costs.

2. If the air is sprayed with water or an open vessel of water is kept near an arrangement the life of the flowers is prolonged.
Explanation:

As discussed in 1 above, this is a question of the relative humidity of the air around the shoots. Loss of water to the air and gain of water from the vase is always going on. Too much loss by the flowers to dry air results in wilting or even withering because the water cannot be replaced quickly enough, even if the stems are stood in water. The surface area of the exposed leaves and flowers is enormously greater than that of the submerged stem in the flower container.

3. Small flowers and seed heads, especially those of dandelion 'clocks' may be fixed with hair lacquer spray.

Explanation:

This gives a rigid outside covering, sticking the parts together and sealing the outer surfaces against water loss. This all helps to prolong life. As hair lacquer is intended for use on people it is less harmful to plants than a spray of varnish might be although this can sometimes be used with success. It is important to spray at the right time to get the best effect. In the case of the dandelion clocks this is obviously before the tiny parachute fruits become too loose on the head. The hairs are all stuck together into one large, interlinked mass which is not so easily dislodged. Naturally it is still delicate but it is certainly not unusable.

4. If the tips of spikes of blooms such as gladiolus or lupin are removed, the spikes can then be placed at angles farther from the vertically upright in a display.

Explanation:

The tips form part of the region which produces the hormones responsible for the response to gravity, as well as assisting in the perception of the force of gravity. If the tips are removed the plants respond less to being held at an angle and do not bend so markedly from their arranged position.

FURTHER READING

Cut Flowers for the House by W. E. Shewell-Cooper (Collins).

Japanese Flower Arrangement by W. Teshigahra (Kodansha International).

The Senior Florist by R. Coleman (Pelham Books).

CHAPTER FOUR

Complex Tips and Their Explanations

By their very nature the hints and ideas which are discussed in this chapter will probably be amongst the most controversial in the whole book. This is because they often have to be used with almost scientific precision to be of benefit, and also the effects may not be visible for some days or even longer after the treatment. It is therefore our opinion that we should spend a few lines discussing the way in which new ideas and suggestions should be tried out.

It is extremely easy to set up an experiment which 'proves' to one's own satisfaction that such and such a treatment has prolonged the life of cut flowers. Somebody gives us an idea and we dash out, collect a few flowers from the garden, and try it out. This time the blooms last a full week. Don't we remember that last year they only lasted a couple of days? It must be a good treatment! The scientist would never accept such evidence. There are so many other influences which could have caused the differences. The flowers might not have been at exactly the same stage of development, the temperature or the air humidity may well have been different, and so on.

To be convinced the scientist likes to see the results from a 'control' sample set up at exactly the same time, using the same flowers at the same stage of development, kept in the same type of vase, in the same room at the same temperature. The only difference must be the actual treatment we are trying to test. A single bloom in each case would not be enough so we should use three or more, carefully chosen to be as close as possible with the flowers in the control vase both with respect to foliage and flowers. A note book is essential to put down the details in. Very few people have the gift of total recall after five minutes let alone five days or five weeks. Photographs and drawings can be a great help. Even a rough drawing forces one to look in detail at the plants rather than just gaining a vague impression of them. Try drawing a picture of any common plant with which you are really familiar and then compare it with the real thing. In many cases it is impossible to finish the drawing because we just can't remember exactly where the leaf joins the stem or just what shape the leaf is. The artist's impression can expose a multitude of unobserved details.

Additives to the water

It may come as a surprise to many flower arrangers to learn that many of the substances which have been passed on by word of mouth as formulas to prolong the life of cut flowers have, in fact, been tested under carfully controlled conditions by trained scientists. The results do not always agree with what we have been told by others. There is commercially great interest in finding ways of keeping cut flowers alive and well for longer than the untreated ones last. Certain firms produce special mixtures which their scientists have produced and tested for them. Naturally the exact composition of these is a closely guarded secret. A few ideas can be found by a casual search through books but those who wish to delve more deeply than the brief account which we are going to give, are warned that they will have a long search both to find, and certainly to understand, why certain chemicals will work and others will not.

Perhaps the easiest way to illustrate this is to discuss in detail the principles behind one particular mixture which has been found to be of genuine value. In the experiments used to find it, well over a hundred different substances were tested either alone or in various combinations.

When plants are growing under natural conditions, they produce their own sugar by the process of photosynthesis which was mentioned in Chapter One. When cut off and placed in the rather dim light inside a room the shoots are not able to do this so well and therefore the shoots tend to become short of sugar. This can be supplied by adding it to the water of the vase so that it can be taken up by the shoots with the water. The ordinary cane sugar which we use in tea and coffee and cooking can be used. Actually this sort of sugar does not enter the cells of plants very readily so that some people prefer to use glucose instead, although the ordinary type of sugar does break down readily into glucose and another sugar which is known as fructose, so that there may not be much difference between using the different types of sugar. About one teaspoonful in a pint of water is all that is needed. Using sugar alone can therefore prolong the life of some flowers and foliage. The sugar however encourages the growth of micro-organisms in the water and they may cause it to smell quickly or produce slime which blocks up the water channels of the shoots and cuts down their water supply. The answer to this is to add another substance which will inhibit the growth of the microbes. All sorts of things can do this, even the old-fashioned penny which contained the metal copper. The soluble form of silver known as silver nitrate can also be used. Only very small amounts of these metals are needed otherwise they may damage the shoot as well as stopping the microbes from growing, so one has to be careful. Another effect of minute trace of metals, especially aluminium salts, is to stabilise the colours in certain flowers so that they do not fade quite so quickly.

Another reason why shoots do not last in water is that they grow older and the flowers pass from young buds, through maturity to old age and senescence. If we could slow down this process, our flowers would last longer. One simple way to do this is to keep the plants cool. When the temperature is raised all changes, both chemical and physical, take place more rapidly, including the attainment of old age—this does not, of course apply to people! Botanists have discovered various chemicals which can slow down plant growth. Some are used as growth retardants, for instance to slow down the growth of grass so that the lawn does not have to be cut so often. If these are placed in water at the right concentration they will hold buds at an early stage of opening and slow down petal drop. If used alone, the colours of flowers may be affected so that another substance, perhaps one of the metals mentioned previously, may have to be added as well.

The complexity of any explanation or simple rules for additives should now be clear. What works for one person may not work for another because their tap water is of a slightly different composition, or they have not mixed the ingredients in the right amounts, or picked their flowers in the same state, or used the same varieties of flowers. Our best advice is to test each idea for yourself, keeping careful records and using controls of untreated but similar shoots.

Forcing foliage and flowers for floral arrangements.

Although under natural conditions many plants are limited to a particular season the flower arranger very often wishes to make the season start earlier, to make it last longer or even to use plants and flowers completely out of their natural season.

There is no simple way of doing this which will work for all flowers at all seasons of the year. Each plant must be considered separately and its particular features considered. There are however, certain basic principles which will give some direction if you do not know the exact treatment for the plant in question. It should be remembered that some plants are just naturally more amenable to manipulation than others. The plants for which there is exact information are usually the ones which have considerable economic importance and have therefore been valuable enough to be used for the elaborate and lengthy experiments which are necessary.

Several external influences affect growth. Water supply is of course one which we have already considered from some aspects. Probably more important from the viewpoint of getting flowers and foliage out of season, are the temperature and the length of the daylight hours to which a plant is subjected. In a temperate climate such as that of Great Britain, the temperature in the winter is lower than that in the summer and also the length of the daylight hours is much shorter in winter. Actually both of these factors show considerable variation, but of the two the hours of daylight can be predicted much more accurately than the temperature.

Many plants are able to control accurately the season in which they flower or become dormant by 'measuring' the daylength. In most cases it is only fairly well developed leaves which can receive the stimulus of light but in the case of a few plants with 'leafless' winter twigs it is the relatively undeveloped leaves present in the buds which do so. Actually botanists have found out that it is really the length of the dark 'night' rather than the light 'day' which really controls flowering in susceptible plants. Except when considering scientific theories to account for flowering, or when using days made up 'artificially' of more or less than twenty-four hours, this is of little importance to the flower arranger! In the most critical cases only one or two days with the right number of hours of light and dark will cause flowering. Other plants are much less influenced by daylength or may even be completely indifferent to it.

In many simple cases we can produce plants and flowers earlier in the year by giving the plant a little warmth and protection. The same methods may also prolong the late summer displays. The perception of the stimulus of low temperatures is important for breaking the dormant condition and stimulating flowering in many plants which come from countries with a cold winter. Usually it takes several weeks for the cold treatment to be effective, which is an advantage to the plants since it means that a chance few days of cold weather at the wrong time of the year will not cause flowering or bud-break too early in the winter. Many plants have extremely complicated requirements for special temperatures and daylengths before they will flower. This is partly because the plants have developed to suit the climate of the country in which they actually originated. Because many plants have been transferred from one country to another and changed by breeding it may take some very skilled detective work to find out what conditions are suitable.

So far we have talked about the plant as if it was simply a passive object receiving information about daylength and temperature from the outside. Actually, of course, it is a living thing and is undergoing internal changes all the time. It is therefore necessary to divide the growth of a plant up into different stages. Each of these stages may need different external conditions to develop to its fullest potential. Thus, although lower temperatures may stimulate the initiation of flowers, these will probably not actually develop until warmer weather stimulates their growth. The experimenter therefore has to find out exactly when the flowers are first formed. It may surprise people to find perfectly formed flowers inside buds and bulbs months before they actually come into flower. Is it during last year's growing season, during the winter or during the current growing season? Clearly, if they are formed before or after the winter there is not much point in varying the winter conditions to get more flowers formed. Of course, other aspects of the growth of the plants may benefit from special treatments.

How does all this complicated discussion affect the flower arranger? What it really means is that we can use certain well-tried methods for a few plants ourselves. If we do not have the facilities to do this (and they can be rather difficult to provide in a normal household) we can buy our material already prepared to flower or actually flowering out of season. The flowers which a florist sells have often been produced in another country and imported because this may well be less expensive than trying to modify the conditions in one's own country by artificial control of temperature and lighting.

Most treatments for bulbs involve rather complicated and accurate control of the temperature so that it is more convenient to buy them ready prepared. Remember, of course, that they will only be 'prepared' for a single season and so will not flower early next year. Selection of suitable cultivars will also give a spread of flowers through a season.

Dormant shoots from trees and shrubs form the easiest material and probably are most used by the ordinary person. Those which flower in the late summer usually form their flowers during the current year's growth and cannot be hurried easily into flower. *Buddleia*, *Fuchsia* and *Hibiscus* all belong to this group. Many other trees form their flowers within dormant buds in the previous season and need the cold temperatures of winter before they open in the spring with the return of warmer weather. Familiar plants of this type are oak (*Quercus sp*), ash (*Fraxinus excelsior*), elm (*Ulmus*

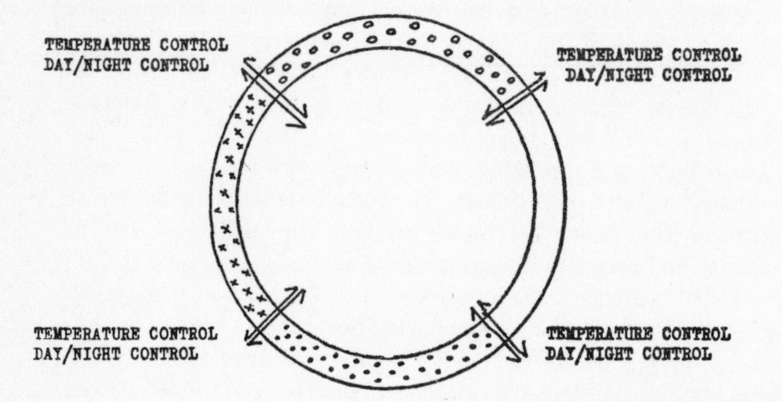

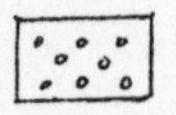

FIG 29a. Cycle of flower formation and flowering periods in perennial plants

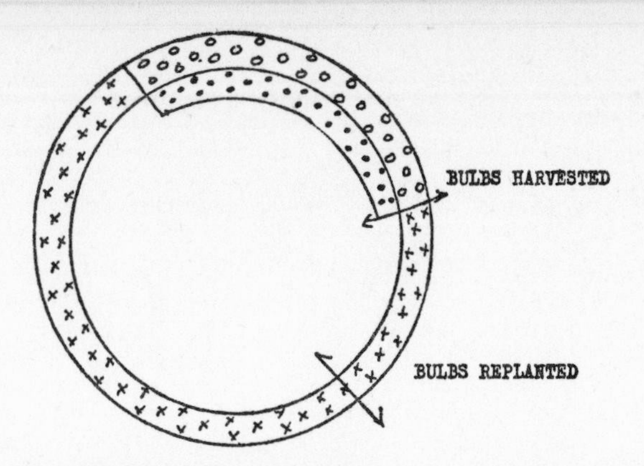

FIG. 29b. Cycle of flowering and flower development in bulbous plants such as *Amaryllis, Hippeastrum* and *Nerine*. Flowers are formed for next year during the current years flowering period. In *Narcissus* the flowers are initiated after the current year's blooming but before the bulbs can be harvested. The period of dormancy may be shortened by higher temperatures but low temperatures are needed for root and leaf growth

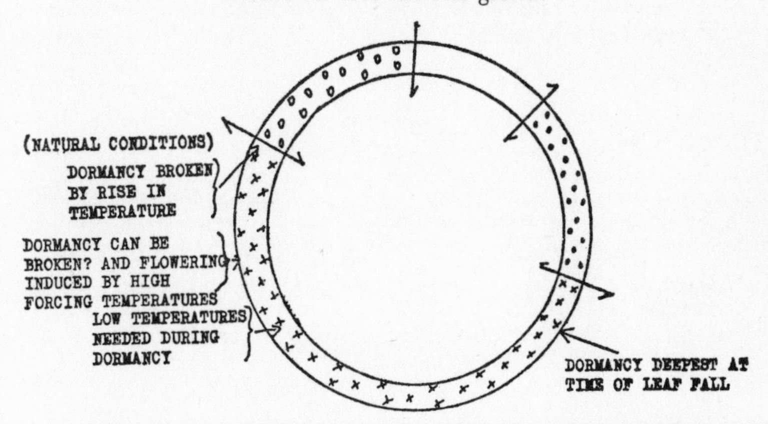

FIG. 29c. Cycle of flowering and flower formation in some deciduous flowering trees. Apple, peach, almond, forsythia and lilac flowers are formed in the previous year. The period of dormancy can only be broken after low temperatures have been experienced. Thus it is useless to bring twigs indoors for forcing until January unless they are refrigerated for various periods of time. Dormancy is usually deepest directly after leaf fall. Twigs of forsythia may be brought in any time after mid-October and placed in a refrigerator at 2°C (28°F), for five weeks. Then soaked well to overcome drying and placed in a bucket of water. If kept at 60°–65°F for 10 days, flowering should start.

Lilac should be defoliated in September–October then chilled at 28°F for 4 weeks then forced at 90°F. After late November no chilling is needed but forcing temperatures are high (110°F). White varieties have proved to be better than mauve because imperfect development of colour appears with forcing

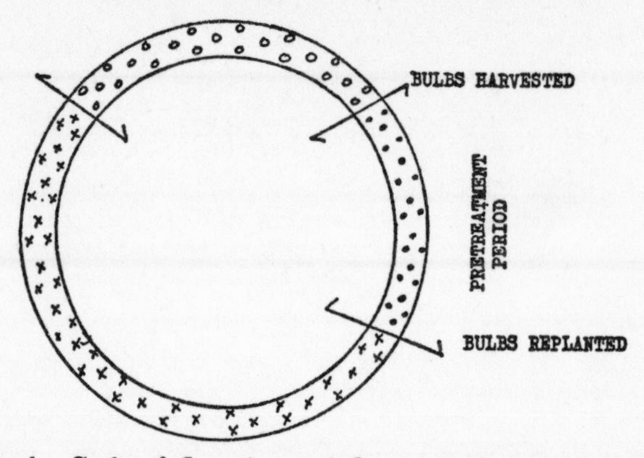

Fig. 29d. Cycle of flowering and flower development in bulbous plants such as hyacinth and tulips. Flower formation takes place after the bulbs are harvested but before replanting. Treatment at this time may be instrumental in producing early flowers. The period of dormancy may also be shortened by 'forcing' in higher temperatures

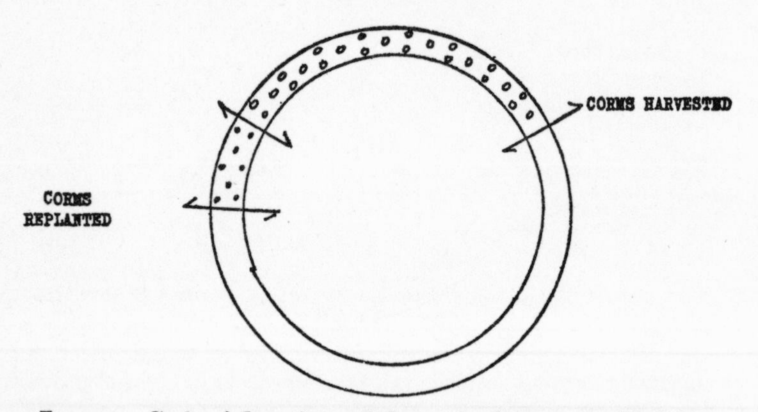

Fig. 29c. Cycle of flowering and flower development in *Gladiolus*. Flowers are initiated after planting of corms and the flowering period follows quite soon. Rest and dormancy lasts from harvesting until replanting

procera), Sycamore (*Acer pseudoplatanus*), apple (*Malus sp*), plum (*Prunus domestica*), peach (*Prunus persica*), pine (*Pinus sp*) and blackcurrant (*Ribes nigrum*). If we bring these shoots indoors before Christmas they will usually fail to open, but early in the new year, when they have had a lengthy period of low temperatures, they will open their buds in two or three weeks. The later they are collected the more rapidly they will open. If you want to force them before they have had the low temperatures outside for several weeks, place them in a refrigerator at about or just above freezing-point (NOT in the ice-box) for four or five weeks before bringing them out into a warm room or heated greenhouse. In certain plants such as hazel (*Corylus avellana*), willow (*Salix alba*), almond (*Prunus amygdalus*), ash (*Fraxinus Excelsior*), oak (*Quercus sp*), elm (*Ulmus procera*), jasmine (*Jasminum nudiflorum*) the flower buds grow at lower temperatures than the leaf buds and so, under natural conditions the flowers will appear before the leaves. This may not always happen when we force the plants artificially.

There are a few plants, notably beech (*Fagus sylvatica*), birch (*Betula alba*) and larch (*Larix europaea*) which open their buds in response to lengthening daylight and these can be opened at any time from leaf-fall onwards by placing the shoots under continuous light from lamps in a warm greenhouse. Under natural conditions, it would be the increasing daylengths of spring which would do this.

Many trees and shrubs in the garden can be forced by similar methods. Forsythia responds quite readily. After it has had the necessary cold treatment, lilac can be forced by giving the shoots a really high temperature (around 110° F.) for about four days and then placing at a more normal temperature of around 60°F. Warm water can be used for the high temperature treatment as also with plants like *Prunus triloba* and *P. serrulata*, which only need about twelve hours at the higher temperature. Water as 110°F. is warm but still not too hot to put one's hand in. Use a thermometer if in any doubt. It will be necessary to replace the water with more at the required temperature as it cools down—obviously not immediately it does so unless you have nothing else to do for a few days!

Plastic vases

Many people displaying flowers in new plastic vases have complained that the flowers wilt prematurely or do not open as freely as buds would normally behave when china or glass vases are used. We had found (using anemones, narcissi and roses) that this is so, although old, well-washed and used plastic containers are quite all right. As many of us know, drinks taken from new plastic beakers have a characteristic 'taste' which is somewhat unpleasant. It seems that some of the chemicals used in the manufacture of plastics are present on the surface of the finished articles and, although soon washed off, have a definite and immediate effect on flowers, causing wilting or preventing the opening of the buds. We are engaged on some research on this subject but are far from being able to make sweeping statements about either all plastics or even one type of plastic on all flowers. We understand that flowers exhibited in shows have been ruined and we therefore think it is worth recommending that all plastic containers be well and truly cleaned with plenty of water before use.

Dyeing fresh flowers

Sometimes it is desirable to include flowers in an arrangement in colours which are not available in those plants offered on the market. It is difficult to explain why these flowers should not be substituted by other cultivars well known to be available in the desired colours. Might it be the perversity of human nature demanding blue carnations or purple chrysanthemums? Whatever it is we offer a few points of interest on this matter.

Certain flowers, notably hydrangeas, can be turned from pink to blue by treating the plants with a form of aluminium salt which may be purchased at garden centres and seed shops. Pink hydrangeas thrive in alkaline soils and the blue hydrangeas are typical of acidic soil. After dissolving the blueing substance in water, this is usually applied several times to the roots, and in due course the well-known large heads of flowers appear. The treatment must be repeated to ensure continued blueing.

Some white flowers respond well to dyeing and certainly last longer coloured. Such plants may have a tendency to turn brown if damaged by rain and are perhaps better coloured. Carnations, narcissi and *Chrysanthemum maximum* are dyed more than any other flowers, the results being startling and judged successful by those needing unusual colours. There are three ways in which dyeing can be carried out.

The most common method is by absoption. The cut stems are stood in a dye solution which rises in the normal conducting channels of the stem until it reaches the petals, causing them to become coloured. Usually the colour remains for about a week but gradually the base of the petals turns white (as the stems are now standing in the normal water given to the arrangement), but the tips of the petals remain coloured for some time.

Cut blooms may also be placed on sheets of paper or in polythene bags and dusted with powdered dye. This is not often used because it is difficult to shake out the excess dye and the dye powder usually ends up over everything.

Flowers may be dipped into a solution of dye (to which a few drops of washing-up liquid have been added to increase the spreading of the dye). The flowers must be left until the required colour is obtained and then removed and allowed to dry. Obviously there must be much experimentation here in timing and strength of the dye.

Dried Flowers, Flower Heads and Seed Heads

Dried parts do not take up water containing dye actively as do fresh, living plants. They need to be dipped or painted with dye. Grasses and glixia (*Syngonathus elegans*) are most often dyed quite brilliant colours.

Many sorts of dyes can be tried out but culinary dyes (often used to colour icing sugar) have proved very successful, probably because they are not toxic to plants and a wide range of colours are available.

It has been recorded that powdered copper sulphate spread around tulip bulbs in the soil changes the colour of the flowers. We have no experience of this and hesitate to recommend it as copper sulphate is known to inhibit growth and spoil the soil.

Compatible plants

One often hears it said that such and such a flower should not be mixed with certain other ones. In some cases the problem is merely one of aesthetics. It may just be that the flower shapes or their colours do not harmonize together. What is very often implied, however, is that when mixed flowers will not last as long as if used separately.

We are sure that many of our readers will be able to give examples from their own experience but we have been unable to find any scientific work which gives a detailed account of this and clearly more information and research into this subject is needed. There are several possible reasons which could account for the effect of one plant upon another. The botanical term for this effect is 'allelopathy', which emphasises the harmful aspects of it.

As mentioned previously, plants may exude sugars and other food materials from the cut ends into the water. This stimulates microbial growth in the water. If a second plant in the arrangement is more easily damaged by the microbes, or its water channels become blocked more rapidly, it may well die sooner than expected.

A second explanation comes from the well-known fact that some living things produce substances which kill or inhibit the growth of other ones. The most obvious examples for the ordinary person are the antibiotics which are produced by micro-organisms and can be used to combat various diseases. Penicillin, which is produced by a fungus, is a familiar example. Many ordinary garden and wild plants also contain substances which are harmful to other living things. Some are poisonous to man or animals, while others can affect the growth of other plants. *Artemisia absinthium*, for example, inhibits the growth of *Lathyrus clymena* and several other plants. Some of the toxic substances have been examined by scientists. Others probably form the basis of the various gardening practices suggested by those people who tell us that such and such a plant should or should not be planted close to some other species of plant. It is quite possible that similar harmful substances could pass from shoot to shoot in a vase via the water.

The leaves of cherry laurel (*Prunus laurocerasus*) and some other plants are known to produce the highly poisonous 'cyanide' when crushed and as this is volatile, smelling like crushed almonds, it is possible for it to spread in the air to other living things. We have already mentioned ethylene gas, which is produced by some plants and can cause other flowers to close or leaves to fall off. Many other examples probably exist, which have not been fully investigated. If you come across any, make careful notes about them and, if possible, carry out the controlled experiment suggested at the beginning of the chapter. Your information could be of use to others.

FURTHER READING

The Encyclopaedia of Floristry by V. Stevenson (Collingridge).
The Control of Growth and Differentiation in Plants by P. F. Wareing and I. D. J. Phillips (Pergamon).
The Flowering Process by F. B. Salisbury (Pergamon).

CHAPTER FIVE

Drying Parts of Plants

The amount of water contained in a fresh plant is quite amazing. It may vary a good deal during the life of the plant and from season to season, but it has been estimated that water may make up as much as 90% of the total weight of many leaves. Water is an important constituent of the living part of the cell and of its walls and is, of course, the liquid in which the salts and other nutrients are passed through the plant. It is therefore an essential ingredient for all living roots, stems, leaves, flowers and fruits. Someone once worked out that apple-trees transpire enough water during the growing season to cover the ground around them to a depth of nearly one foot had it all been collected and that an acre of beech-trees transpires tens of thousands of gallons in a growing season. Thus the movement of water through the plant, and the maintenance of a turgid condition to allow this movement, is essential to life.

When the normal growing period of any part of a plant comes to an end its water supply is upset and the water loss by evaporation is not replaced. Eventually it shrivels and dies.

Knowledge of the structure of the plant part (be it flower, leaf or fruit), and of the manner of its water loss may contribute to an understanding of its appearance when dead. Different sorts of plants respond to the changing conditions very differently. Finely branching, but strong and rigid, vascular tissue which supports the thin tissues between plus a gradual loss of the water in the plant usually results in an undistorted product of great beauty. Once, on seeing a long herbaceous border at the gardens of the Royal Horticultural Society at Wisley (Surrey), which had been left, uncharacteristically, unattended far into the winter, the tremendous value of the shape and colour of 'dead' herbaceous plants was really appreciated. It made one realise that the great autumn clear-up which goes on in most gardens in the cause of tidyness, deprives the garden of seemingly moonlit extravaganzas which come to life with a touch of frost or a shaft of winter sunlight. Flower arrangers can capture this special beauty in their dried flowers, but for them, the preparation of dried displays needs special care so that 'deadness' may be avoided.

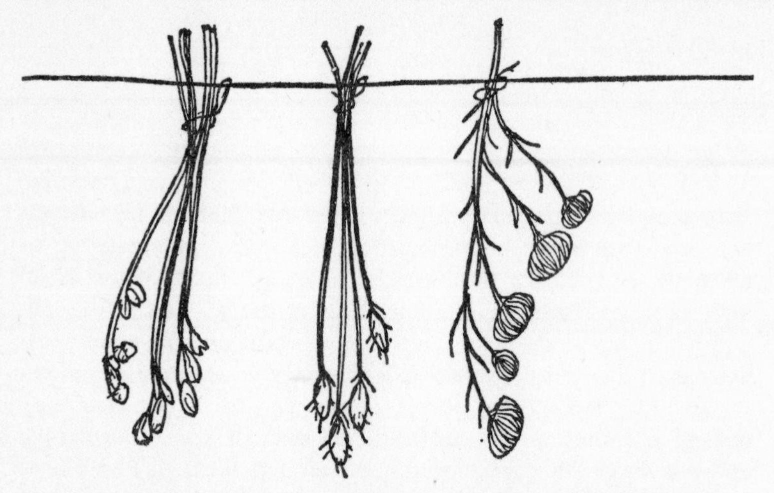

Fig. 30. Small bunches of flowers may be tied up and hung on a string in a clean, airy shed to dry. In this way straight, strong stalks are preserved

Flower heads and leaves which will dry naturally, need only to be collected at the right time and protected from the blackening influence of autumn mists and rains. The exact time for picking is difficult to predict and a good deal of trial and error must be expected as no two autumns are the same. If the individual stems are tied into small bunches and hung head downwards in a reasonably clean, airy garage or shed, the result is a straight shoot and clean dried flower-heads suitable for almost everlasting use. If the only place available for drying is full of cobwebs and dust, large polythene bags perforated with many holes may be used over the flowers. Sunlight tends to bleach the plant pigments and therefore near-darkness is desirable for preservation of bright colours. If the flower stalks become brittle when dried and fail to hold the heads it is better if the heads are detached at an early stage and laid flat on a newspaper to dry. They can be wired onto false stalks when needed. A frame of chicken wire or a cake tray, allowing the air to circulate freely all round the heads can also be used. Most grasses dry very well (see Plate 5).

If almost any part of a plant is collected before it is dead and placed with its stems in either a solution of one part glycerine to one part water, boiled together and cooled or in a solution of one part antifreeze to one part of water thoroughly mixed, a more supple, less brittle everlasting shoot results when the shoot has absorbed these materials. These solutions are taken into the stems, pass through the xylem vessels and into every stem, leaf and flower. When drying begins only the water evaporates and the glycerine or antifreeze remains in almost every part of the plant, producing suppleness and often attractive colours which together are of value in winter arrangements. Leaves of *Fatsia japonica*, and fig, branches of beech and oak become deeply coloured in rich browns and reds after such treatment in glycerine or antifreeze mixtures.

All foliage to be preserved should be in perfect condition. Leaves which are mis-shapen, show holes, or have parts eaten away look poor when preserved. Smaller sprays (18 inches in length), are more successful than larger branches as the penetration of the mixture is easier and quicker than in large branches. It is also essential that the leafy shoots are not left on the tree too long. Once the natural autumn tints are showing it is too late to start artificial ways of keeping the leaves attached to the stems and of encouraging subtle colours.

The reason for this is probably obvious to the reader. Transpiration (evaporation) from the leaves is the force which draws water into the plant. If the leaves are not still in a reasonable state of health the water and whatever is in it will never reach the top of the shoot. The exact time of cutting is difficult to predict but early autumn is the time to watch for starting the treatment. These treated branches will retain their leaves and can be used over and over again. They may even be pegged out on the clothes line in the rain for cleaning because the glycerine or antifreeze will have penetrated right inside the tiny cells and will not be easily washed out.

Under natural conditions, the leaves of deciduous trees fall off at the end of the growing season. Actually the leaves of evergreens also eventually fall off at some time in the year but since this happens over a longer period of time it is not so noticeable. This natural loss of leaves is totally different from what happens when we cut leaves off with scissors, leaving an unprotected wound. In nature the plant has been preparing for the loss of its leaves for several weeks. Certain minerals and foods have been withdrawn from the leaves and a layer of protective cork has been growing across the base of the leaf stalk.

To the outside of this is a layer of very weak tissue which eventually disintegrates letting the leaf fall off but leaving behind a fully protected leaf scar. This process is known to botanists as abscission and is controlled by certain plant hormones. If we pick our leaves too late in the season, when they are already grown, the abscission layers will already be formed and even if we succeed in getting glycerine or antifreeze into the leaves the leaf bases are already weakened to such an extent that they will easily fall off (see Plate 3).

Table I

Foliage suitable for preservation in glycerine or antifreeze solutions (see text)

Name of plant	*Appearance when dried*	*Notes*
Aspidistra (also called Parlour palm) (*Aspidistra elatior*)	Rich red-brown	Very slow, takes from 4–6 weeks
Atriplex (*Atriplex halimus*)	Deep red	Very quick, less than a week
Bay (*Laurus nobilis*)	Shiny chocolate brown	Rather slow, leave for 4–5 weeks in solution before using
Beech (*Fagus sylvatica*)	Rich brown, nut cases remain attached	Must be gathered before natural colour turns. Often left too late
Begonia Rex	Metallic brown	Fairly quick. Leaf stalks may be rather weak
Bracken (*Pteridium aquilinum*)	Light brown	Small lengths are best. If underdone, the edges will curl in a warm room
Cherry laurel (*Prunus laurocerasus*)	Shiny deep brown	Slow. Shoots with up to five leaves best
Choisya (*Choisya ternata*)	Shiny deep brown	Rather slow. Leave for 4–5 weeks before using. Attractive citrous smell even when dried
Cotoneaster (*Cotoneaster horizontalis*)	Tiny leaves turn red-brown	Short lengths of the flattened stems are best. Must be cut before leaves begin to turn naturally
Elaeagnus (*Elaeagnus pungens*)	The brilliant splashed fresh leaves turn a disappointing uniform brown	Slow, 4–6 weeks

Eucalyptus (Gum) (*Eucalyptus gunnii*) etc.	Pale brown	When shoots are cut hard back to retain juvenile foliage, the prunings may be preservd in glycerine at any time of the year. Beautiful shapes and colour
Fatsia (*Fatsia japonica*)	Large deeply lobed, chocolate brown	Single leaves treated stalks may need support as the leaves are so big
False aralia (*Dizygotheca elegantissima*)	Narrow palmate leaves, chocolate brown	Leaf stalks are weak and need wire support
Ferns, various	Pale brown	Rather slow, short lengths are best
Ivy (*Hedera helix*)	Light or dark brown	Very short lengths or single leaves are best. Quick, 1 week
Magnolia (*Magnolia grandiflora*)	Dark brown	Single leaves. Slow, 3–4 weeks. Must be cut while still green
Mahonia (*Mahonia japonica*)	Large, shiny dark-brown pinnate leaves	Very slow. 4–6 weeks
Maidenhair fern (*Adiantum sp*)	Pale brown and very dainty	Not easy, but worth trying
Megaesia (*Bergenia cordifolia*)	Large, rounded and dark brown	Single leaves are best. Rather slow, 3–4 weeks. Stalks may need support
Oak (*Quercus sp*)	Pale or red-brown	Short lengths. Acorns remain attached if cut early
Pear (*Pyrus communis*)	Red-brown	Must be cut when young. Quick, 1–2 weeks
Pittosporum (*Pittosporum tenuifolium*)	Pale brown, with wavy margins	Tends to drop if cut too late
Rhododendron (*Rhododendron sp*)	Red-brown	Slow, deepens in colour with keeping. Single leaves do well
Solomon's seal (*Polygonatum odoratum*)	Arching stems with pointed oval leaves, light brown	Must be picked early before natural colour fades. Difficult but worth the trouble
Sweet chestnut (*Castanea sativa*)	Pale brown	Short lengths or single leaves.
Spotted laurel (*Aucuba japonica*)	Deep brown, shiny	Rather slow, 3–4 weeks

Although we have mentioned the green pigments in leaves we have said little about the other colours. In fact green leaves also contain other yellowish pigments (known to the scientist as carotenoids) which we cannot usually see unless the chlorophyll is absent. Thus in the golden cultivars such as golden privet which contain only small amounts of chlorophyll these yellow pigments are obvious to the eye. The leaves of flowering plants grown entirely in the dark are yellow because light is needed for the final stages in the production of chlorophyll. We can also see these yellow colours in autumn leaves because the chlorophylls are more rapidly destroyed than the yellow pigments when the leaves start to die.

The first sign of autumn in many leaves is therefore the change to a yellow colour. Even the yellow pigments become altered, however, so that different shades of yellow develop. In addition some leaves develop totally new pigments of a different type which were not previously present in the leaves. These for example are the bright reds of various trees, shrubs and climbers.

When completely dead the ordinary green leaf which has yellowed will usually become brown. Some of the brighter colours may be more persistent and still remain, although in rather duller shades when the leaves are completely dead and dry.

What of the glycerine or antifreeze treated leaves? The changes which take place are basically similar to those in normal, untreated leaves. Although the treatment 'preserves' the leaves it does not keep them alive. In fact the living processes in the cells are upset sooner than they would have been so that, for example, the abscission layers do not form and the leaves remain attached to the twigs. Also other changes can take place which would not normally occur under natural conditions. This means that new pigments and colours can develop. Because the glycerine or antifreeze prevents the complete drying-out of the leaves they do not become dry and brittle as would normally happen when they die. The preservative effect of the chemicals also stops the leaves from going mouldy as they usually would if they were kept wet with water alone. This preservative action of solutions with a very high osmotic strength (i.e. water with a lot of another substance dissolved in it) is the same one that causes jam to keep. Thin, watery, fruit juices made from the same fruit will not keep unless completely sterilised and kept sealed.

If Nature does not dry plants to our satisfaction, we must try more drastic measures. Small leafy shoots of horse chestnut and beech may be pressed between sheets of newspaper and ironed with a slightly warm iron. The water content of the leaves is reduced without the parts shrivelling. Even fronds may respond to the same treatment but often curl at the edges when arranged in a warm room. Grasses may be treated in the same way. For the best results, however, most plant material is dried more gently by pressing between sheets of absorbent paper placed under the carpet or weighted down by a pile of books. The more tidy minded may prefer to buy or make up a plant press as used by professional botanists (see Plate 6).

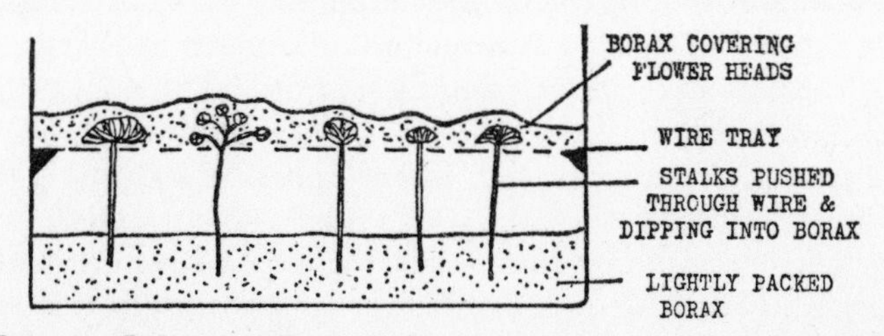

FIG. 31. It is not difficult to fix a wire tray in a box and spread a piece of gauze on it. Borax is shaken into the bottom of the box and flower stalks are pushed through the gauze and the wire and the heads lightly covered with powdered borax. The whole thing can be slipped into a polythene bag and placed in a dark cupboard. When perfectly dry the flowers can be shaken free of borax which can be used over and over again

The flowers should be opened out, the side shoots pruned to prevent overlapping and the plant placed in tissue paper (called 'flimsies') which may be moved from one used newspaper folder to a new bone-dry one without further disturbing the delicate flowers and leaves.

Although pressing gives good results for grasses, sedges and rushes it is obviously not much use for fleshy, succulent plants which lose their colour and shape while being flattened and dried. In fact this is the way in which harbarium sheets are prepared for museums and great reference places such as Botanic Gardens where the naming of plants is studied. Properly mounted and named herbarium sheets last for many years and from the reference material for studies of the identification and nomenclature of plants (see Plate 7).

Many large flowers and flower heads retain their colour and dry well if picked just before they are fully open, placed in a box and packed with dry powdered borax or with silica gel. Both of these can be obtained from the chemist. The silica gel is usually sold in tins to be used for drying out the moisture between the panes of glass in double glazing systems. Both can be redried in a warm (not hot) oven for us again. Alternatively the flowers may be gathered and hung upside down in polythene bags containing the powder and shaken gently so that the powder comes in contact with the flowers. Borax and silica gel both extract water gently from flowers without the damage to the pigments which heat would cause. Scabious, roses, some dahlias and chrysanthemums may be treated in this way but there is a good deal of trial and error because the time of picking is very difficult to predict. Hydrangeas are best left as long as possible on the plant to ensure maturity and the full development of strengthening tissues before cutting and then placing the cut end of the stem in 2 inches of water. After this has been absorbed leave them to dry out naturally.

For flower heads to be dried satisfactorily their structure must be understood. Flower heads of the Compositae are often used in dried flower arrangements and they all have a structure similar to that of the familiar daisies and dandelions. Each of the petal-like structures is really an entire flower, ready, if given the opportunity, to set seed. The seeds are often dispersed by feather 'parachutes'. These are not of much use in flower decorations because, with the warmth of the room, dispersal takes place, covering everything with flying parachute fluff. Surrounding the mass of individual flowers, however, there may be a cup-like collection of small, rather leafy, bracts. This is known to botanists as the involucre. In many members of the daisy family it is of a papery texture, 'everlasting' when dried and often brightly coloured. Thus in the preparation of some composite flower head such as helichrysums and xeranthemums, the central florets may be rubbed off, blowing gentle puffs of air until only the everlasting bracts remain. Alternatively the heads can be picked when the bracts have not opened fully and still enclose the florets.

Bullrushes are made up similarly of tiny florets which ripen to form feathery fruits. They are usually picked under-ripe and may last several months before a slight tap or movement starts the release of thousands of tiny plumed fruits. To prevent this happening the heads may be sprayed with hair lacquer. The billowing mess of untreated heads which have burst is a sight to behold . . . and to clear up (see Plate 8).

FIG. 32. Flower head and longitudinal section of the everlasting flower of Immortelle, *Helichrysum bracteatum*. It should be picked before the central papery bracts have become fully opened and the true florets are still enclosed. If left until older the central florets set seed and disperse fluffy fruits

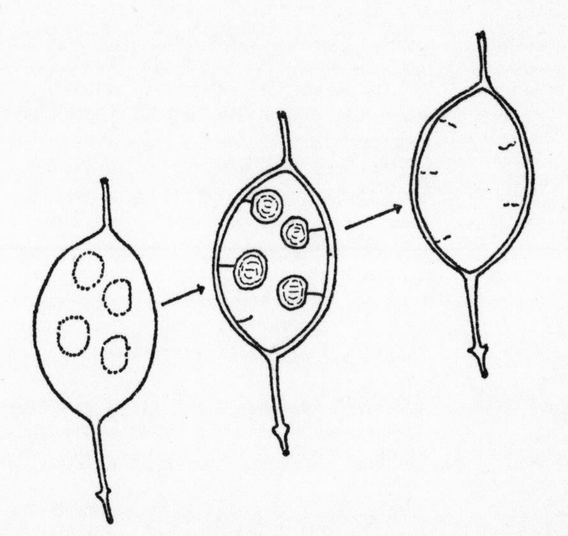

FIG. 33. Removal of the carpel walls and seeds from the pods of honesty (*Lunaria sp.*)

The two papery carpels which form initially the pods of honesty (*Lunaria sp.*) are usually removed as the heads dry releasing the dark, flattened seeds (which germinate very easily) and revealing the thin, silvery-white, septum so beloved in its natural beauty or titivated with glitter at Christmas.

Seed heads of teasel, poppy, montbretia, *Crocosmia*, *Acanthus*, sorrel, *Trifolium stellatum* and *Aconitum* are usually collected when natural drying has just started and are bunched into small groups hanging upside down, pegged to a line in a well-ventilated dry shed. Seed heads are valuable material because of their natural appearance. It is natural that dehiscent (splitting) seed boxes release their seeds while still attached to the parent plant.

Table II

Flower heads and seed heads to be dried naturally. In most cases treatment with glycerine or antifreeze solutions makes the material too supple for large arrangements.

Name	*Notes*
Achillea (*Achillea filipendulina*)	Cut when brightest yellow. Dries well in Borax.
Bears breeches (*Acanthus spinosus*)	Cut when the top buds are opening. Good for large arrangements, and will keep for almost ever.
Bells of Ireland (*Molucella laevis*)	Cut while still in flower and hang upside down. (Plants should be well staked otherwise stalks are bent and difficult to use). The bell-shaped calyx remains green if dried in the dark. May also be skeletonised, well worth the effort.
Cardoon (*Cynara cardunculus*)	Very large flower heads on strong stems. The leaves need to be removed as they dry badly.
Clary (*Salvia horminum*)	Careful drying in borax or by hanging upside down in small bunches usually preserves the mauve or pink colour of the bracts.
Cupids dart (*Catananche caerucea*)	Cut in August and hang upside down to dry. Mauve colour usually persists well.
Delphinium	Cut when top bud is just opening. Smaller spikes dry best and keep their colour better than talls.
Globe amaranth (*Gomphrena globosa*)	Cut just before fully open; hang up to dry.
Globe artichoke (*Cynara scolymus*)	The smaller, less spiny (and edible) edition of cardoon. The heads dry best when cut just before they open. The large thistle-like mauve heads make excellent huge arrangements when fresh and dry quite well even when open.

Name	*Notes*
Globe thistle (*Echinops ritro*)	Cut when quite mature then the round blue heads keep their colour well.
Golden rod (*Solidago canadensis*)	Pick when at brightest yellow. If too old, turns brown; if too young, small flowers drop off.
Gypsophila (*Gypsophila panniculata*)	Cut when 80% of the flowers are open. Hang up to dry. The tiny flowers dry up but still remain white.
Honesty (*Lunaria annua*)	May be left 'growing' indefinitely. Hang up to dry. The two penny-shaped carpels may be gently removed scattering the seeds. Will keep for almost ever.
Hydrangea (*Hydrangea macrophylla*) hortensia forms	Cut when fully open or leave until papery. Place in two inches of water and when this is used up leave to dry. The large heads of sterile flowers last for long periods when dry. Cultivar 'General Vicomtesse de Vibraye' is excellent rosy-red and may be turned a good blue if watered with aluminium salts while growing.
Immortelles (*Acroclinium*, *Helichrysum*, *Helipterum*, *Rhodanthe*, *Xeranthemum*)	(or everlasting flowers) are flower heads surrounded with papery bracts which dry well and retain their brilliant colours. All should be cut with long stalks and hung up to dry. The thin 'neck' beneath the flower is the weakest point and flower heads are sometimes detached and dried separately. All should be cut before the true flowers in the centre of the head are fully developed.
Statice (*Limmonium sinuatum*)	Mauve, yellow or pink varieties dry probably better than any other plant although the green winged stems become rather grey. Hang up to dry. The wild *L. vulgare* found near sea shores is excellent for dried arrangements and is finely branched.
Ling (*Calluna vulgaris*)	Cut when barely fully out and handle as little as possible. Retains its natural colour well.
Love-lies-bleeding (*Amaranthus caudatus*)	Dries in borax best. The green variety 'viridis' is worth growing.
Love-in-the-mist (*Nigella damascena*)	Cut only when the large seed capsules are formed and dry slowly. The curious shape and pale brown colour are useful attributes.
Monkshood (*Aconitum napellus*)	Small shoots dry best; cut when top bud is just opening.
Oriental poppy (*Papaver orientale*)	Cut seed heads when fully ripe and hang up to dry. Shake the seeds out.
Ornamental grasses	Grasses dry naturally very well. Cut in small bunches and hang up. May also be pressed without loss of shape. All should be cut just before they are fully ripe.

Name	*Notes*
Protea (*Protea longiflora*) etc.	The S. African flowers dry very well naturally and last a long time. Good brown colour without looking dead.
Pearly evrlasting (*Anaphalis sp*)	Cut before the central florets open. The small heads dry well if hung up.
Sea holly (*Eryngium maritimum*)	Cut only when quite mature. The stiff branches, spiny leaves and bluish heads must be dried slowly.
Shoo-fly-plant (*Nicandra physalodes*)	These tall annuals with forked stems and persistent calyces dry naturally and make interesting arrangements. Hang up singly to dry.
Scabious (*Scabiosa atropurpurea*)	Pincushion heads on long stalks need to be hung up to dry Colour is poorly retained but the shapes are useful.
Teasel (*Dipsacus fullonum*)	Stiff forked stems and conical heads with spinous bracts should not be cut until fully ripe and after all the mauve flowers are over. Hang up. One of the best plants for drying.
Zinnia (*Zinnia elegans*)	Double flowers are best; cut when flowers are just fully open. Hang to dry, slowly.

Male catkins of sweet chestnut fortunately fall to the ground from the tall trees soon after midsummer and may be dried and even dyed in culinary dyes. They are curious, fluffy, pipe-cleaner-like structures, often as much as eight inches long. They are very useful in small, dried-flower arrangements.

Stems of Chinese lantern (Physalis franchettii) drop their leaves in early autumn. The tiny tomato-like fruits shrivel but the splendid extended, lantern-like calyces surrounding them, retain their bright orange colour for a long time. The best way to preserve and dry these is simply to place them in a vase with no water and use as required. Shoo-fly (*Nicandra physalodes*) are treated similarly and can be painted or 'glittered' at Christmas.

Ornamental gourds, custard marrows, pomegranates and rosehips are protected by hard impervious outer coats and dry naturally and last a long time. Their life and cleanliness is enhanced by a coat of varnish applied thinly after the fruits are dried. The central tissue usually dries completely making the fruit very light in weight. All these are indehiscent (non-splitting) fruits which have no special method of dispersing their seeds and rely mostly on birds to spread them. The birds are attracted by their brightly coloured outer skins, succulent interiors and many seeds.

Acorn cups are composed of many small bracts fused together at the base of each flower and growing to accommodate the acorn which eventually dries and is shed. Some species of oak, especially the Turkey oak produce rough, spinous, cups which dry well and are attractive in dried flower pictures.

Table III

Large single fruits to be dried naturally

Name	*Notes*
Chinese lantern (*Physalis franchettii*)	Wait until the leaves have dropped before cutting. Hang up to dry.
Corn (*Zea mays*)	Ornamental varieties produce coloured grains. The cobs should be left as long as possible on the plants, then cut off and dried slowly (in airing cupboard) without removing the large papery leaves which dry pale yellow and are very ornamental. If the cobs are picked too soon they shrivel.
Custard marrow (*Cucurbita sp*)	Flat, fluted fruits must be hard and ripe before they are picked otherwise they will not keep.
Gourds (*Cucurbita pepo*)	Fancy shapes and colours dry naturally if picked when fully ripe. They may be varnished.

Many other fruits such as pomegranates, small oranges, capsicums, nuts, acorns and acorn cups last a long time if kept dry.

Cones of many species of *Pinus*, Douglas Fir (*Pseudotsuga*) and Larch (*Larix*) dry naturally but if warmed in the oven remain 'open' permanently. Cones of Cedar (*Cedrus*) and of fir (*Abies*) tend to disintegrate with ripeness but can be used when young.

Beech cupules split into flower-like formations allowing the dispersal of the nutlet and when ripe are a rich brown. They need no further attention before they can be used in various decorations. They can be collected earlier when still green and attached to the leafy shoots. Then they can be allowed to take up glycerine in the usual way to avoid any danger of them falling from the stems. The tiny beech nut cases of *Fagus sylvatica laciniata* are especially dainty.

Cones and twigs bearing cones (pine, spruce, larch and the small cone-like flowers (female) of alder) dry naturally but tend to close in damp weather. If dried in a slightly warm oven the open position of the scales is fixed more permanently. Most cones are of an attractive brown colour which is permanent if they are kept clean. Cones may be dried in boxes of dry sand and then stored in dry cardboard boxes (see Plate 9).

Cones of the true fir (*Abies procera*) and of cedars (*Cedrus* spp.) tend to disintegrate into their component scales on drying but are beautiful if used fresh and young. They are rarely shed in this whole state however and as they are borne high up on the trees are difficult to obtain. In addition they tend to be rather heavy for most displays. Those who have gathered cones will be aware that some of them exude a very sticky resin with a characteristic smell. To some this smell is extremely pleasant but to others it may prove objectionable. If you really must use these cones it is often possible to get rid of some of the excess resin by wiping it off with methylated spirit but this therapy is rarely permanent.

Skeleton leaves are a special sort of dried material which can be very useful in floristry. The skeletons must be perfect and unbroken and the greatest care must be exercised in their preparation. Skeleton leaves are those in which only the veins remain, the soft parts having been removed. This occurs naturally when leaves fall to the ground and worms, snails and soil micro-organisms eat or etch away the softer parts of the leaf, leaving the tougher veins which sometimes become further whitened by weathering. All too often the natural result is imperfect due to a tear. Since skeletonising is essentially a natural process we can sometimes get Nature to do the job for us by putting our selected leaves into a tub of water stood outside and inspecting the progress of decay after a few weeks. As this process is rather slow and can sometimes be a little smelly, the more impatient among us will wish to obtain results more rapidly. With suitable leaves the soft parts may be removed by immersing the leaf in a pan of water to which a handful of washing soda has been added. Take care not to use an aluminium pan as this is corroded by washing soda and will be ruined for other purposes. Simmer for about one and a half hours and leave to cool. The exact time to use will depend partly on the actual leaves which are being used so that it is probably best to get the time right using some trial leaves. The next stage is to hold the leaf under a gently running cold tap and scrape very carefully with the fingers or a blunt knife, taking especial care to avoid damage to the veins. The leaf skeleton will usually emerge as a dull brown colour from this treatment but can be left in a weak solution of bleach (one teaspoonful to a pint of water) overnight. Then rinse thoroughly and dry between sheets of soft kitchen paper under a light weight. Handle with care. When the skeletons have been pressed flat (in a book or under weights) they may be coloured, using a few drops of culinary dye (the type used for icing sugar) or other dye. They can also be sprayed with hair spray or glue and immediately dipped into glitter. They can even be sprayed or dipped into a paint. Only certain leaves are suitable for this treatment of skeletonising. These are the ones with strong, woody veins such as magnolia, holly and laurel. Chinese lanterns are shoofly plants and *Molucella* (Bells of Ireland) produce inflated calyces which often skeletonise naturally.

FURTHER READING

Dried Flowers for Decoration by V. Stevenson (Collingridge).
Garden Foliage for Flower Arrangement by S. C. Emberton (Faber and Faber).

CHAPTER SIX

Winter Arrangements and Display Aids

When the first frosts have robbed the garden of the few remaining summer flowers, the attention of the flower arranger turns towards other materials. Several problems conspire to make this a difficult time. Firstly, of course, there is the sheer lack of variety in the available flowers. Gone for another season are the extravagant displays of warmer days. Most flowers will now have to be purchased and financial considerations will usually limit the flowers used and the numbers of them which are available. Secondly there is the problem of making the flowers last in an arrangement. This is made difficult because conditions in the home will now be much more unfavourable for plants.

If the central heating is on it will almost certainly be too dry for the plants to grow normally. It may even be too hot, for the cool rooms of summer will be subjected to more accurate temperature control. If the heating is gas or oil, there may be the danger of fumes affecting the flowers although with correctly adjusted burners this should not be a problem. More serious will be the lack of light. The light which enters from outside will be much reduced in intensity because of the low angle of the sun in the sky and the presence of more cloud. The actual hours of daylight will also be much reduced. All this means that the green parts of the plants will be much less able to produce food by photosynthesis. The warm room temperatures will aggravate the situation by making the plants use up what reserves they do have more quickly.

Suppose that we do not have central heating. The position is still not very good. Light will be equally poor but on top of this there will tend to be a very severe variation in temperature. High day and evening temperatures will encourage the plants to use up their reserves and lose water to the atmosphere. Water loss will probably be greatest where electrical heating is used since most heaters which burn a fuel both produce some water in the process and also encourage the drawing-in of fresh, moister air into the room through cracks and crevices. Once the heating is turned off temperatures may fall rapidly and in many rooms may almost reach freezing—only to be boosted again the next day. All this hardly makes an ideal situation for delicate blooms which have been cossetted previously in a carefully controlled greenhouse and are probably flowering long before their rightful time.

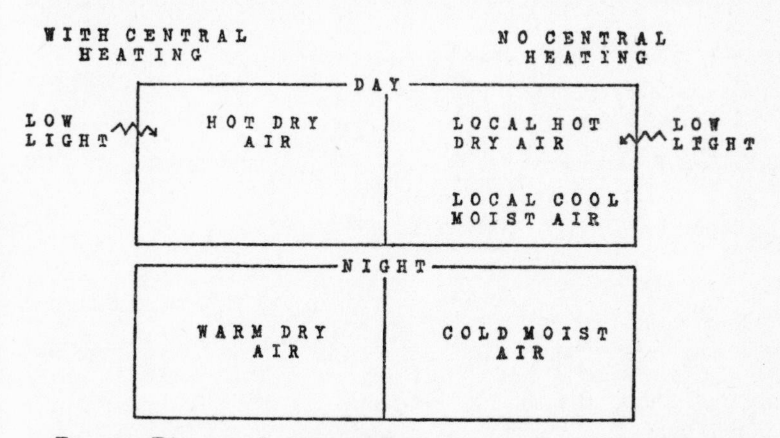

FIG. 34. Diagram of winter conditions in a room. A comparison of conditions

Many people give up the unequal struggle and settle for an entirely dried arrangement (see Plate 10). In this there is nothing to be ashamed of. Unfortunately, some normally ardent arrangers seem to look upon winter as the close season and having once made up a display will leave it alone for months on end. Of course no one would dream of doing this with living flowers so why do it with dead ones? The eye becomes tired of the same display and it becomes dustier and dustier. In winter we need to ring the changes regularly just as in summer. To do this one must keep a plentiful and varied supply of dried material. In Chapter Five we gave some hints on drying your own material. If that has not produced sufficient then a very modest outlay each week will soon build up a really fascinating collection from the exotic materials which many shops stock. Grasses, of course, are available in profusion, but one should keep an eye open for the more unusual stems and fruits which crop up from time to time. Buy them when they take your fancy for they may well be gone when you come back next time (Plate 5). Naturally you will not want all the materials on display at once so find a suitable large box, suitcase or old chest of drawers for storage. Do not just throw in the delicate specimens and push it shut. Gather the dried plants up carefully. If you have many different kinds, gently roll up each sort separately in paper. Now place them into a large air-tight polythene bag and close it securely. This will prevent insects and moisture getting at the plants. Of course you must make sure

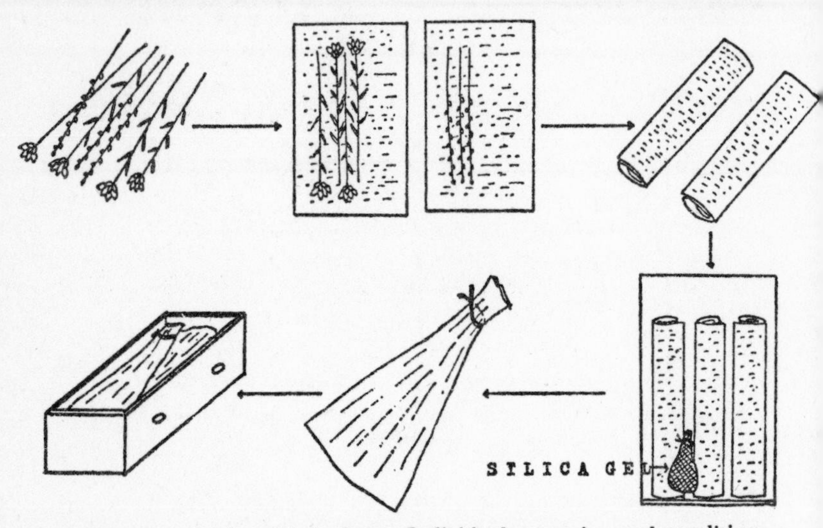

FIG. 35. Storing dried plants. Individual stems in good condition are rolled in paper, and stored in a closed polythene bag in which a muslin bag of silica gel keeps the air dry

that all the specimens are fully dry before putting them into the bag otherwise when you open the bag later you will be greeted by a cloud of musty spores from mould growth. A muslin bag of dried silica gel inside the polythene one will help to remove any traces of moisture which remain in the air. Keep the polythene bag well out of the reach of small children at all times because the properties which make it suitable for the storage of plants also make it lethal to children if they should put it over their heads while you are out of the room.

The more adventurous arrangers will combine dried and fresh materials. This is easily done using a second container, or wrapping the bases of the dried material in waterproof foil or polythene. The inclusion of some living material may help to remove the 'dead' look

Plate 9. Some dried female cones which show a wide variety of shapes and sizes.

Plate 10. A selection of dried garden and wild seed heads and twigs.

Plate 11. Prepared wooden bases lent by Trevor Sears of Wormley, Surrey and photographed by Muddle of Godalming.

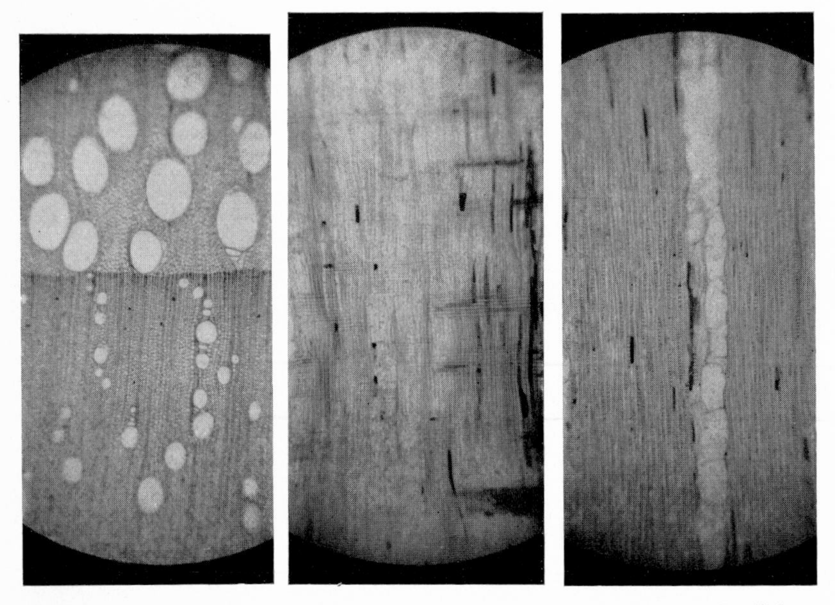

Plate 12. Wood under the microscope. *Left to right:* Transverse, Radial longitudinal and tangental longitudinal sections of ring porous wood (compare with Fig. 41). (Photographs lent by Margaret McKendrick.)

of dried material but must be used with sympathy or the result will be incongruous.

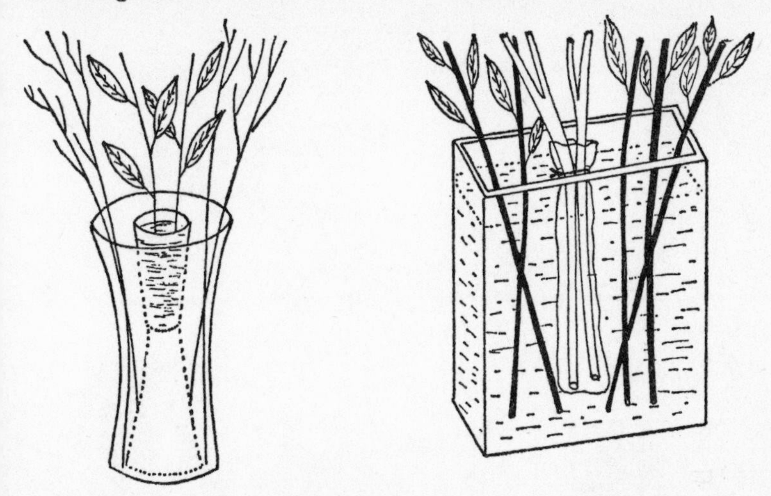

FIG. 36. Suggestions for mixing dried and fresh material in arrangements

Mention of buying dried plants brings our attention immediately to the question of varnishing, tinting, painting and artificial flowers. Most flower-arrangers will only work with natural material but where do we draw the line? Obviously the items we collect ourselves are natural and the cheap, shabby plastic replicas are artificial. Many of the better plastic flowers, however, can only be seen as such when you approach closely, and the many shops which sell them testify to their popularity with someone. What is most distasteful about them to the student of Nature is their lack of subtlety—nothing pressed in a mould can ever bear the individuality of a natural object.

Varnishing has already been mentioned as an aid to the preservation of some fruits and perhaps we can allow this but what of dyeing and staining? Many of the dried grasses and flowers which can be bought are artificially coloured and it must be left to the good taste of the individual to select or reject the various shades which are available. Some mingle well with untreated natural materials while others clash offensively.

It is at Christmas that we really come up against the problem of painting and artificiality. Is a piece of butcher's broom (*Ruscus aculeatus*) which has been sprayed silver still natural? Many flower

FIG. 37. Foliage available in winter. (A) Holly (*Ilex cornuta*), (B) Holly (*Ilex aquifolium* 'Ferox'). (C) Ivy (*Hedera helix*). (D) flowering leaf of Ivy. (E) Yew (*Taxus baccata*). (F) Western red cedar (*Thuja plicata*). (G) Barberry (*Berberis sp.*) (H) *Skimmia japonica* female plant. (I) Laurestinus (*Viburnum tinus*)

arrangers will say no and use only such fresh material as they can obtain—there is plenty of good foliage about if we care to look. Good taste with regard to dyeing and staining cannot really be defined—it either exists or it does not!

The same problem does not exist with regard to display aids. These are objects to set off an arrangement and as such we can treat them in whatever way we please and integrate them into the design to any extent. They are best considered in turn.

Wooden Bases

Wood may be used in a great variety of ways in flower arrangements. To consider it as only an adjunct to arrangements does less than justice to it since, even if only used as a base plinth, it still forms an integral part of any artistic arrangement. A wooden base can take many forms which vary from the rigidly formal to the appealingly simple.

With any wood it is essential that it is thoroughly seasoned otherwise the warmth of a house causes unsightly cracks to appear, especially along the 'rays' of the wood. Seasoning originally meant that trees were cut down and the trunks or prepared parts of them were left to dry out slowly in the open air, with perhaps only slight protection from the elements throughout several seasons. This takes a considerable amount of time, but since the process is slow the wood does not tend to twist or crack due to strains being set up during the drying.

Much of the wood which can be bought now is kiln-dried, which really means that the natural drying has been speeded up under very carefully controlled conditions. If the drying is too rapid, as often happens if we saw down a tree in the garden and cut out a few thin pieces for use in the house, the result is disastrous. When we cut down a tree, most of the wood is already dead but some parts, particularly the living ray cells which radiate out like the spokes of a wheel have a high water content. When these dry out, they shrink more than the bulk of the wood and the result is a crack. It should be obvious that young, sappy wood is more likely to crack than more mature pieces or those which have been cut for some time and have slowly been drying out.

Really fine polishes are usually not very practical on bases because they tend to mark rather easily with water or get scratched by the bases of vases. Once damaged, they cannot easily be returned to the original condition. Pieces of wood of a range of species can be bought from shops or timber yards, and each has its own distinctive appearance and properties. Other pieces may be left over from home handiwork and can be salvaged for use. There is a wide range of veneered boards also available which can be used to advantage.

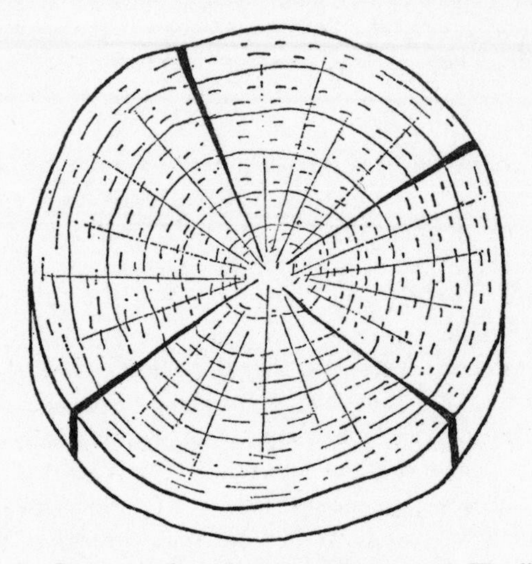

FIG. 38. Diagram of cracked unseasoned wood. The line of weakness is along the rays

Softwood is the term used commercially to describe the timber of coniferous trees such as pine, spruce, fir and cypress having needle- or scale-like leaves and usually being evergreen.

Hardwood timber is that of broadleaved trees such as oak, elm, maple, willow and ash. These trees are usually deciduous.

Although the terms are generally accurate the timber of some softwoods such as yew is harder than that of lime or willow.

Wooden bases for floral arrangements are usually cut obliquely from trunks or branches and chosen from timber showing interesting colour or grain (Plate 11). The bark may or may not be left in place; firm, fissured bark adding much to the appearance of the base. Colour differences of heartwood and sapwood are seen in many woods, notably in yew (*Taxus baccata*), laburnum (*Laburnum anagyroides*), false acacia (*Robinia pseudoacacia*), and oak (*Quercus sp*) where the heartwood is dark, very hard and takes polish well, and in many of the softwoods where the heartwood is often yellowish or pale-coloured sometimes exuding resin if it is not fully seasoned. Close-grained woods such as oak (*Quercus sp*), apple (*Malus sp*), and cherry (*Prunus sp*) make good bases whereas holly (*Ilex aquifolium*) is a white sappy wood with greenish bark, splitting along the rays and warping. More valuable woods such as walnut

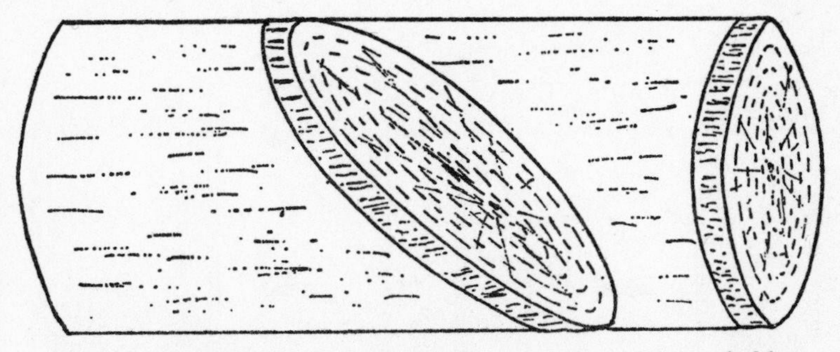

FIG. 39. Diagram to show the oblique cut along which is suitable for bases

with satiny texture, wavy grain and interesting knots are hard to come by but worth seeking out. Many imported tropical woods are available nowadays, some having rich red or yellow colourings which may be intensified and preserved by careful polishing. One particular feature of interest in timber is the knot. This consists of the remains of a side branch which is hard and dark. This is the round piece which sometimes falls out. Of more value to the user of the wood are the attractive markings or figuring which surround the knot.

It is worth while backing a wooden base with felt or foam latex to prevent it scratching furniture on which the arrangement may be displayed.

The collection of wooden bases is quite a hobby in itself, some arrangers possessing a collection of shapes and sizes for all occasions, and, like most hobbies, it inspires further interest and a thirst for more information about the origin, habitat and structure of trees.

When a suitable piece of wood has been obtained it can be quickly treated with linseed oil and then polished with furniture polish. This brings out the grain beautifully. Often more to hand is clear polyurethane or other varnish which renders the wood impervious and unmarked by the damp bases of flower containers. The wood can also, of course, be stained to various shades before, or with the varnishing. The 'grain' of wood reflects the underlying structure of the timber. The most common features which can be seen are short, straight marks due to the presence of rays that are especially seen in veneers or the rings of minute holes found in some transverse sections across stems and branches. These represent the presence of

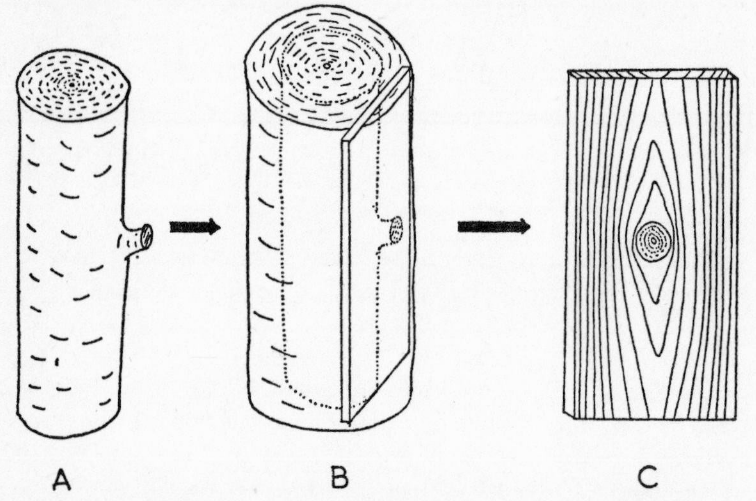

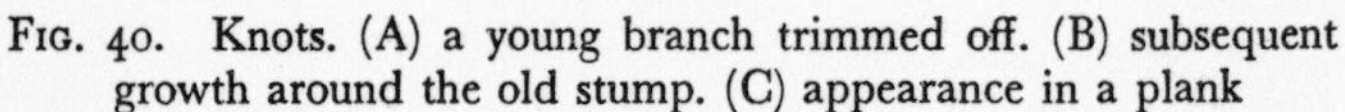

Fig. 40. Knots. (A) a young branch trimmed off. (B) subsequent growth around the old stump. (C) appearance in a plank

conducting vessels of large diameter. Sometimes these are scattered throughout the wood—so-called diffuse porous wood—while in other trees they may be produced only during the spring and then give rise to very obvious annual rings because of the very abrupt change from one season of growth to the next. This second type of wood is known as ring porous. Thus it can be seen that woods are not equally suitable for use as bases (or for use in other ways) so that if you or your neighbour has occasion to cut down a tree with really useful wood make sure that you do not waste it! (see Plate 12).

Bark

Bark is dead, corky tissue, usually formed as a protective cylinder around stems. The interesting feature of bark is that new material

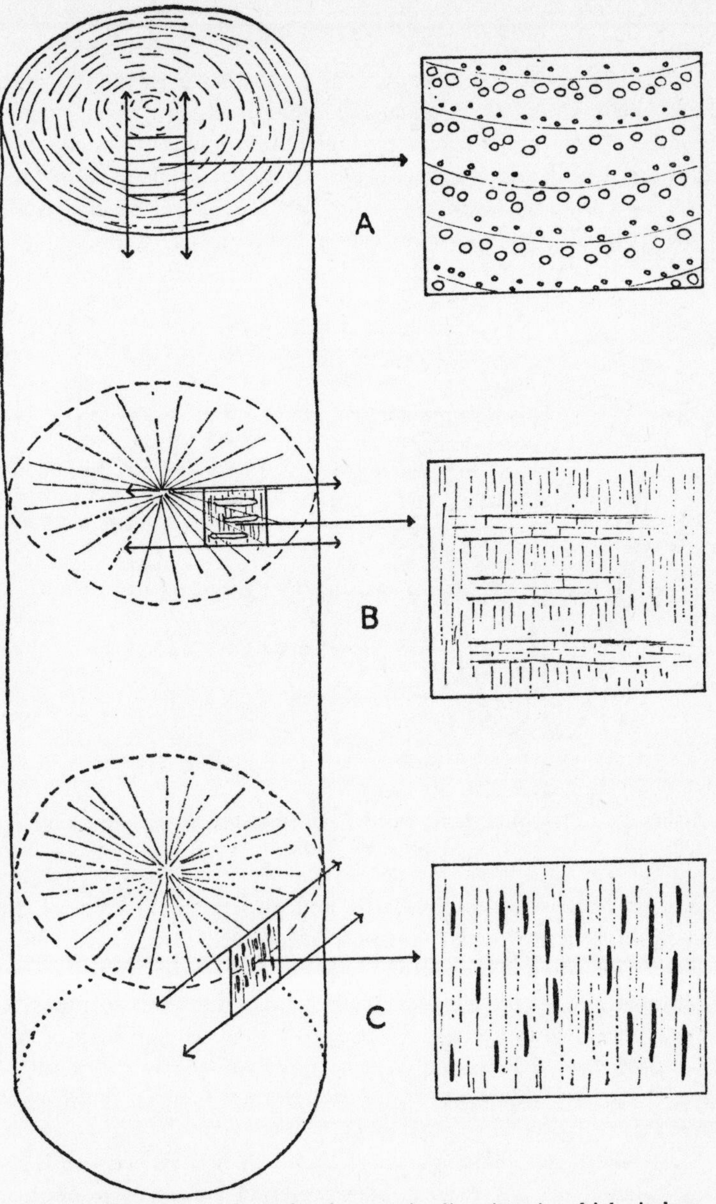

FIG. 41. Diagram to show the three main directions in which timber can be cut. (A) the transverse cut shows the growth rings of the xylem. (B) the radial longitudinal cut (along a radius from the centre) shows the side view of the vascular rays. (C) the tangental longitudinal cut (at right angles to a radius) shows the ends of the rays. Veneers used on furniture are this type. (See also Plate 12.)

is being continually produced, usually on the inside of the old, from special cells which form a cork cambium. As the outside areas become cut off from their water and nutrient supply, tannins and resins are deposited in the bark, giving it a characteristic colour and scent. The ordinary bottle cork is formed in this way on the trunks of the cork oak (*Quercus suber*) where the bark is very thick.

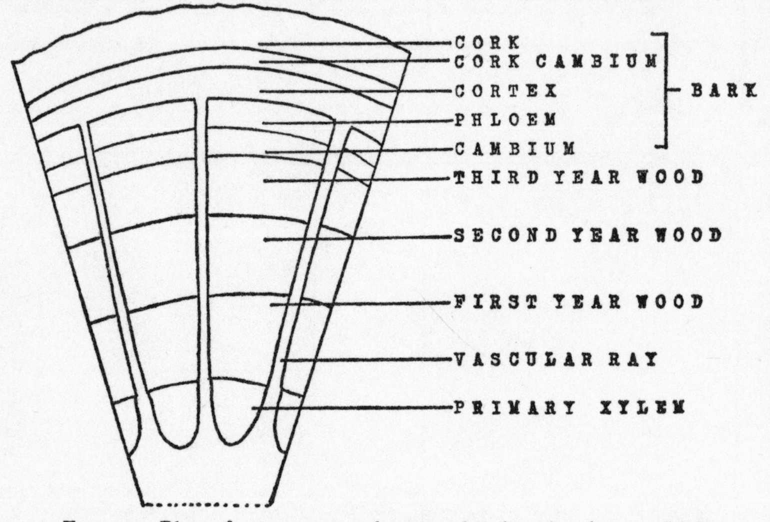

FIG. 42. Plan of a young woody stem showing the tissues which form the bark

Bark is only very slightly elastic and growth in thickness of the tree, in some cases, causes the bark to crack and fissure in interesting and specific ways. The peeling bark of silver birch is caused by short-lived and successive rings of cork being formed and ruptured by ever-increasing internal growth of the stem. Rough bark of oak and apple, and the fissured bark of pines are partly the result of hard, thick bark coming to terms with annual internal growth (see Plate 13).

Many shops sell choice pieces of bark which have been well and truly cleaned up but readers may prefer to collect some of their own. Almost any wood will provide some pieces but really good ones will need a lot of hunting down. Where trees are being felled is often a productive spot to look. When found, the bark will almost certainly be wet, crawling with insects and other creatures and perhaps providing a base for various wood rotting fungi. Have a suitable container—a polythene bag is quite good—into which the selected pieces can be put. Naturally it will not be any good if it crumbles at the slightest touch. Remove what loose material you can whilst in the woods. When you get home the task is to trim it to size and clean off loose dirt, animals and so on. A good rub with a stiff brush may help, perhaps even using clean water from a hose. Let the excess water drain off and dry out the bark in the sun. Failing this, use a gently warm oven which will also help to kill off any mould growing below the surface or the eggs and grubs of insects which cannot easily be removed by brushing. Bark is often used without any further treatment, but it can be varnished if desired. One coat will give some preservative effect, but a clear grain-filler would probably be needed if you ever wanted a shine on it!

Drift Wood

All round the coast and large bodies of water can be found pieces of wood which have been rubbed smooth by the action of wind and water. Often the sun has also bleached them to a pale grey colour. If you buy driftwood from a shop it may well have been further improved by sand blasting to remove any blemishes and decaying areas, and further chemically bleached. It may even be linseed oiled or varnished. Occasionally you may find similar pieces in the forest but these, as well as any pieces you find on the shores, will need preparation before they can be used. The first stage is to remove dirt and loose pieces and give the material a good cleaning. A surface bleaching using dilute household bleach may improve the appearance. (Protect your hands with rubber gloves). The bleach should be thoroughly washed off with clean water before drying in the open air or gently in the oven. If the finished pieces do not stand correctly they can be screwed and glued onto a matching or contrasting base. Make sure that the piece is well balanced, if necessary adding hidden weights underneath, otherwise it will prove a constant irritation and may well damage furniture if it falls over.

Stems, branches and roots

Strictly, driftwood should perhaps come into this category, but it is intended here to include the more natural branches which are still covered with bark and perhaps a variety of other things. These can be collected in fields, woods and gardens and it is always wise to keep a look-out for unusual pieces (Plate 14). Old vine stems, often showing curious shapes may be incorporated into displays. Other interesting stems are those of corkscrew willow (*Salix matsuanda contorta*), hazel (*Corylus avellana contorta*), and the rather rare yew (*Taxus baccata contorta*). Other stems may show interesting leaf scars as seen in horse chestnut (*Aesculus hippocastanum*) or have coloured barks. There will be a demand for both thick woody branches and for slender, more flexible ones.

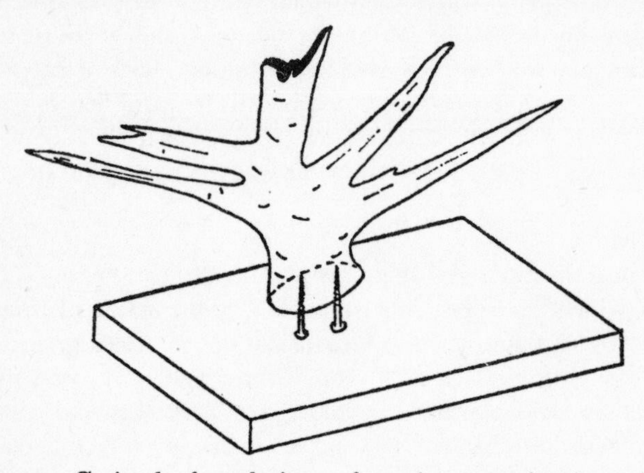

FIG. 43. Curiously shaped pieces of wood may need to be securely fixed to a base

These branches need little treatment except a general trim and clean-up and then gentle drying. As discussed in Chapter Four, it is possible in some cases to force dormant shoots brought in from outside to put out their leaves or even flowers under the influence of heat or light indoors. In this way spring can be made to come early in the same way that treated bulbs can be made to flower at Christmastime.

The stem bases which arise from the creeping underground stems of the giant reed (*Arundo donax*), a familiar sight along the shores of the Mediterranean, and a source of fishing-rods, roofing-canes and other strong, straight articles may be used in an interesting way for decorations. Unfortunately, these reed bases are not found in Great Britain or other places with similar cooler climates but might be looked for when on holiday or visiting warmer places. (Customs regulations may be a problem in some places!) The stems are hollow and extremely woody and should be sawn off about six inches above the ground—they are too strong and woody to merely cut off and something more than a penknife or secateurs is recommended. When the strong fibrous roots have been trimmed off an attractive, somewhat rustic object is made. The hollow stems can be used as candle holders and the branched rhizome forms a firm base.

It is possible to buy in some shops, or to obtain direct if one is very fortunate, pieces of root which have been dug out of the ground, split into suitable pieces and then dried and brushed clean. They can even be stained or varnished. The roots of spanish chestnut (*Castanea sativa*) are particularly suitable for this purpose. To the uninformed, these pieces of root look like stems but are harder and heavier. They are sometimes found under the name of natural wood sculpture and, of course, choice pieces command a high price (see Plate 15).

Carved Wood

If Nature does not provide quite the material which we need, it is only a short step to assist her. With something roughly to shape, we can prune off unwanted pieces and it is then but another step to a more massive removal of wood. The many different types of 'shaping' tools which are now available can make this easy even for the non-carpenter. Staining and varnishing are easily done and the whole subject shades from natural abstract to formal carving as an accompaniment to the flower arrangement.

Moss

Moss is a form of plant life which does not produce flowers. It is spread by vegetative means or by producing masses of spores in special capsules. In size, mosses range from minute to several inches in length, and in form from simple unbranched plants to trailing, much-branched ones. Many mosses take up water readily and are often not much damaged by drying out. They may retain their natural green colour and do not turn brown on drying like many higher plants. One of the most familiar water-holding types is Sphagnum moss which can be found in some damp boggy areas or bought from shops. This is mentioned again later on.

Some of the unbranched mosses form very smooth, velvety cushions and these are very useful for covering over the surface in certain informal displays. Single blooms of early spring flowers, snowdrops (*Galanthus nivalis*), winter aconites (*Eranthis hyemalis*), and primulas (*Primula sp*) can be arranged to advantage in bright green, damp moss in shallow dishes.

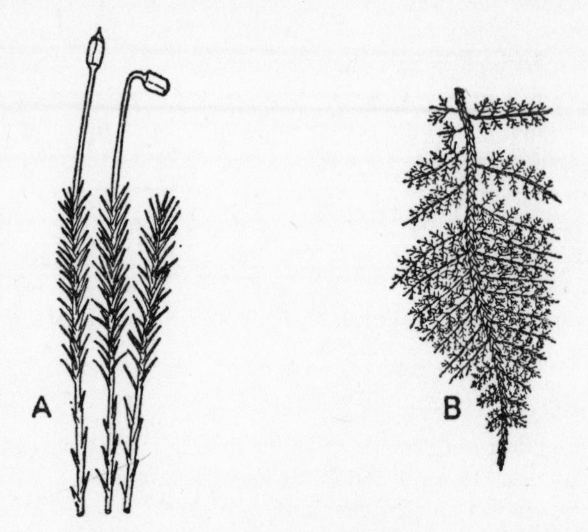

Fig. 44. Drawings of two mosses. (A) *Polytrichum commune*, with straight leafy growth and terminal capsules, this moss forming tufts. (B) a portion of a highly branched prostrate moss, *Thuidium tamariscinum*

Much-branched mosses can be used in a variety of ways to hide stems and give extra appeal to mixed arrangements. Because they do not possess conducting channels in the stems there is no need to place the cut ends in water and they will be quite successful with a gentle surface spray of water. If they are allowed to dry out this will usually result only in a temporary shrivelling which can be easily reversed by giving the moss a little water.

Lichens

Some lichens are superficially similar to mosses but upon even a limited investigation it will be quite clear that they have a very different texture which is usually somewhat stiff and leathery. Lichens are most interesting botanically because they are in fact mixed plants made up of an algae and a fungus forming an intimate association and living together for mutual benefit. To the flower arranger they will be seen mostly as interestingly coloured and textured 'scales' on bark and stones or as small, rather greyish-green, branching growths. Sometimes there is the vivid contrast of bright red spots which are actually masses of the minute reproductive structures which spread the lichens from one place to another.

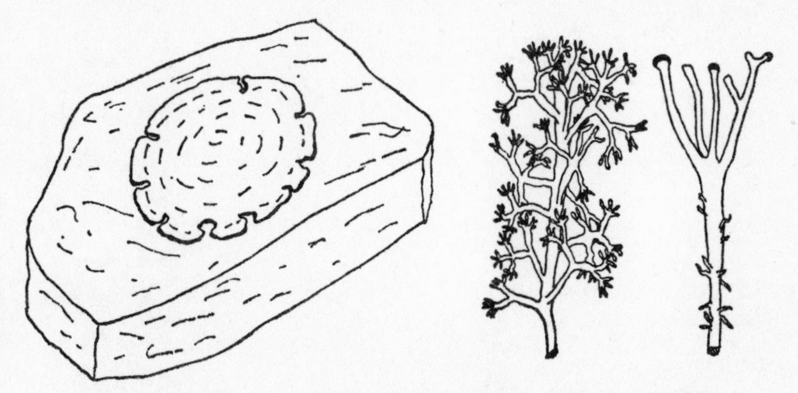

FIG. 45. Lichens of different growth habits; encrusting, highly branched and erect

Lichens can be particularly useful for display because of their ability to be subjected to severe drying without going brown or losing their shape. Once again they have no conducting channels and should preferably be kept out of water.

Fungi

Particularly in the autumn months it is possible to find large numbers of fungi or 'toadstools' in the woods. These can be picked and used fresh if you are able to find one which is not infested with hundreds of small maggots and insects. These find the succulent toadstool a very palatable treat. Some of the toadstools, however, are very poisonous to humans, and all should be treated by the untrained as potentially harmful and not left where young children might be tempted to take a bite from them. Unfortunately the common and very attractive fly agaric (*Amanita muscaria*) which has a brilliant red cap flecked with white is very poisonous, despite the fact that it is frequently used to illustrate children's story books.

Most of the toadstools which are found cannot be easily dried or preserved but many of the woody 'bracket' fungi which grow out from branches and trees can be dried off in a warm oven and will then keep indefinitely. They make an attractive addition to mixed displays.

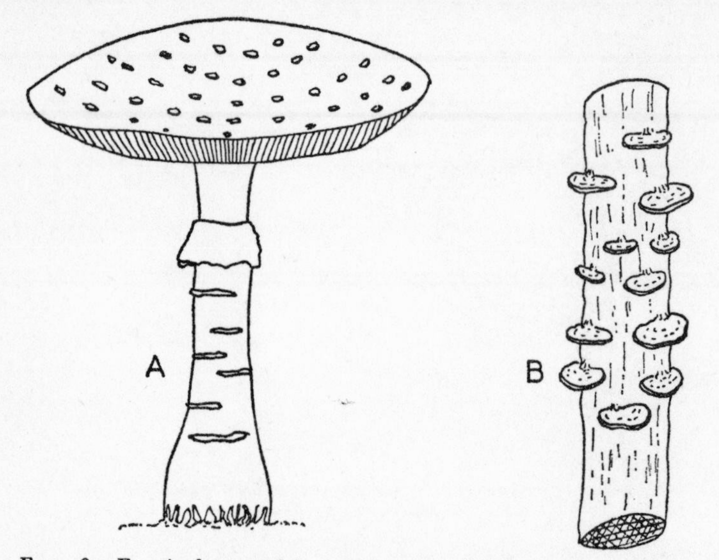

Fig. 46. Fungi often used in arrangement. (A) the poisonous toadstool *Amanita muscaria* which dries poorly. (B) a bracket fungus which dries successfully

Stones

There are two ways in which stones may be used. Firstly small chippings or pebbles may be used to cover bare areas at the base of displays in shallow open containers. These may be collected by oneself on the shore or elsewhere. The salt must be thoroughly washed off of stones from the shore before use. Coloured chippings are probably best bought either from a garden sundriesman or from the aquarist shop. The type of stone gives scope to the arranger from hard granite chips through a range of natural materials to 'synthetic' rock chippings.

Large stones or roughly carved stone can be made an integral part of some displays with colour and texture complimentary or contrasting with the floral or vegetative display. Finer carvings will often form a centrepiece of considerable visual impact if suitably matched in the display.

Stones need little attention other than an occasional wash off with soapy water. If the type of stone is amenable, a hammer and chisel can be used to improve the shaping somewhat but the flower arranger should not become too engrossed in this work unless they still have time for another interest!

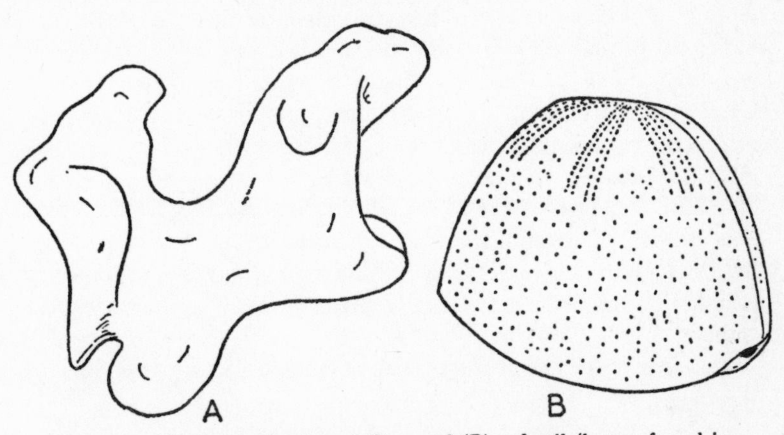

FIG. 47. (A) Curiously shaped flint and (B) a fossil (better found in chalk) which are suitable for some arrangements

Containers

The selection of the correct container is vital to any flower arrangement and those who take the art seriously will accumulate a selection which can be used as required. The shape of a selected container is something rather personal and we will content ourselves with a few general points.

Shallow open containers will have a large surface area of exposed water and this has two main effects. Firstly, evaporation will be greater than from a narrow-necked container and hence the water level will have to be checked more often. Secondly, the oxygen in the water is more easily replaced as it is used up by the submerged parts of the flowers. Water will tend to become smelly more quickly in the narrow-necked container probably because of the lack of oxygen caused by the relatively small surface area of water which is exposed to the air.

Vases and containers of different materials will vary in the extent to which water soaks into them and through them onto the table. The porous ones, such as unglazed pottery, will need a waterproof inner lining or a second container.

Although clear glass containers may reveal attractive stems below the water level, they often cause problems because stains must be completely removed if the visual appearance is not to be spoilt. 'Hard' water can cause unsightly rings to form at the level of the water surface. These can be removed with a good scrub or a little vinegar. The more stubborn, brownish stains at the base of the container require more prolonged treatment with hot water containing washing soda or such other remedies as experience shows to work. Clear plastic containers need to be more gently treated when cleaned or permanent scratch marks may be made which spoil the appearance and provide crevices in which stains form. Some special effects of plastic containers have already been discussed in Chapter Four. The growth of green algae will not usually be a problem if the habit of thoroughly cleaning a vase before putting it away is adopted. If you leave it to clean later and forget it, you must expect trouble.

Dark-coloured and opaque vessels should be cleaned just as thoroughly.

Stem Supports

These may take the form of pin holders or glass blocks with holes in them which stand at the bottom of a vase. These are very useful but not suitable for all flowers. Many stems just will not fix to pin holders satisfactorily. Many arrangers use the synthetic water-holding materials which allow the stems to be pushed into them easily and are particularly useful for shallow displays.

Crumpled wire netting, preferably with a wide mesh is also frequently used, but it obviously needs care in handling if nasty scratches on fingers are to be avoided. Plastic coated wire netting may be used and this is often conveniently sold in small lengths at flower shops and garden centres.

A whole range of plant materials can also be used to support stems in the desired positions. Sphagnum moss holds water well and is virtually free from any micro-organisms so that it keeps very well. It lives naturally more or less submerged so that it does not decay quickly. It can be squeezed dry, rather like a sponge between use.

Reasonably stiff, wiry stems from many plants can be cut and used as packing but they may show or encourage decay quite rapidly because of the many substances which they release into the water and which can be used as food by micro-organisms. An interesting study this (the effects of one plant on another) but little precise information is available.

Plasticine is another frequently used support for the more woody stems but because it blocks the entry of water at the very tip of the stem the plant must be able to take in water through the stem higher up. In this use the plasticine acts rather like wax which is used to stop up the ends of some flower stems to prevent leakage.

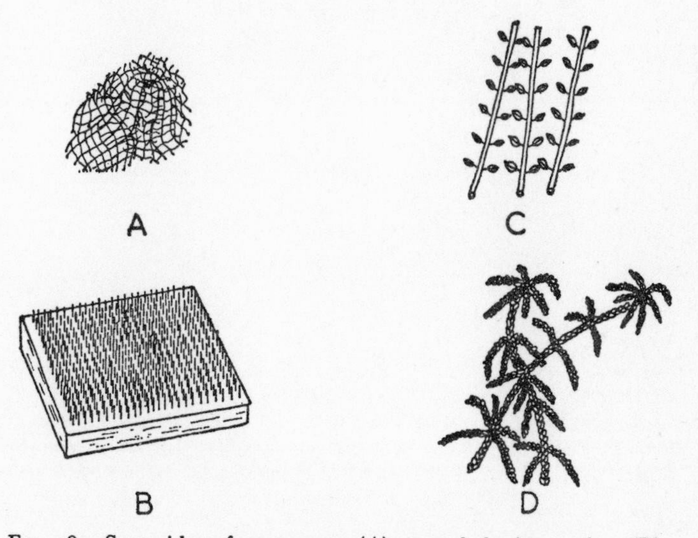

FIG. 48. Some ideas for supports. (A) crumpled wire netting. (B) a pin holder. (C) twigs of *Lonicera nitida*, short lengths used for packing in the vase. (D) Sphagnum; a useful mossy base

Supports for winter arrangements of more woody and dried stems can be made from large potatoes cut in half and laid cut side down on a suitable flat plate or something similar. Twigs of holly for example last a long time when supported like this. Immersion of the potato under water will not be beneficial as its soft tissue will start to rot and soften in a day or so and potato is therefore of no use with fresh flowers.

Making up New Flowers

Although Nature provides a large range of plants which we can use in arrangements, some people like to make up their own new ones. This can be done quite easily with some dried materials. A suitable framework branch is first selected and onto this can be pinned, glued or wired various fruits, seeds, dried leaves or other items which have come from other plants. Used with discretion some attractive products can result.

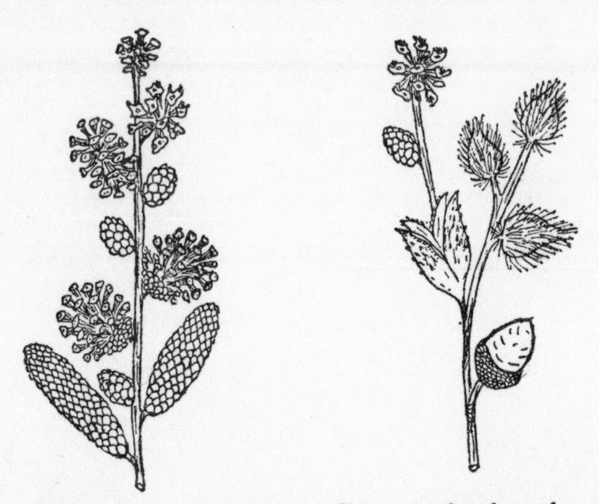

FIG. 49. Constructing new plants. Cones and other short-stalked dried materials can be made more useful to the flower arranger by sticking, pinning or wiring them onto a suitable branched 'stem' so that they can be mixed in with dried and preserved leaves and flowers in the vase. If necessary the joins can be hidden by a little florist's tape

Christmas Displays

The many ideas suggested above can be combined to give attractive and yet economic displays at Christmas in the Northern Hemisphere. A traditional base of yule log of wood such as silver birch (*Betula verrucosa*) will have one or two holes drilled for candle and smaller ones for the insertion of individual dried stems or supports for trailing foliage. Larger cavities in the log can accommodate container for water for fresh flowers. Of course many flowers can be obtained from the shops at this time for those who prefer live flowers.

FURTHER READING

The Driftwood Book by M. E. Thompson (D. Van. Nostrand).
Guide to British Hardwoods by W. B. R. Laidlaw (Leonard Hill).
British Ferns and Mosses by P. Taylor (Eyre & Spottiswoode).

CHAPTER SEVEN

Plant Names

Suppose you go into the garden and pick one of each flower growing there. If they are then laid out on the table, it is at once obvious that some are very similar whilst others are very different. In an attempt to classify them they could be sorted out according to their size, but this would not be very satisfactory because it is well known that the conditions of growth can markedly affect the size of a plant. Of the many different characters which could be used botanists have decided to utilise mostly the details of the flower as a means of grouping plants together. This is because the number and arrangement of the floral parts tend to be relatively stable and hence serve as a good reference point.

Usually the ordinary or common name applies to all plants of a special kind. These names arose from the simple need to recognise medicinal, food or other useful plants and, rather later, decorative ones too. Needless to say, these names are different in all parts and languages of the world. Pansies have about fifty common names in English, and as many in Spanish, French and German plus names in Russian, Chinese, Japanese and Hebrew. Thus although common names are interesting (Chinese sacred lily), amusing (Husband-come-home-however-late-you-be), or picturesque (Flower of the west wind) there is no doubt that they are scientifically unsatisfactory and that when interest in botany became widespread, they caused much confusion. Either plants had too many names or they were not common enough to be named at all. By this time it had become necessary to help Man's tidy mind by gathering similar plants together in groups.

Scientific names are often derived from the common names used in Greece by Theophrastus, a student of Aristotle, who named and classified over 400 plants in his 'Historia Plantarum' around the year 300 B.C., or from those used in the Roman Empire and referred to by Pliny and other Roman authors. After the fall of the Roman Empire, Italian, French, Spanish and Portuguese languages arose largely from Latin. Like all living languages they were always changing and becoming enriched with new words and meanings. Only a 'dead' language, which is not spoken, is stable and precise. Latin became the language of the learned, which usually meant the Church. Botany then, as now, was concerned with feeding the hungry and healing the sick and was the handmaiden of medicine and agriculture. Plants have probably been used in religious rites since before written history. Even we still have our harvest festivals.

Thus it was that plant names became long latin descriptions of how a plant looked and where it grew and how it was used. Clearly a cumbersome method.

Carolus Linnaeus, in the book *Species Plantarum*, published in 1753 reduced the descriptions of all the known plants to a binomial or 'two word name'. A generic name referred to the genus (closely related group) to which the plant belonged, e.g. *Lilium* a lily, and a specific name, e.g. *tigrinum* is specific to and descriptive of the tiger lily. These two names are used throughout the world for tiger lilies and for no other plant. The two words of the binomial always agree in latin gender although this may be somewhat obscure to those who have not studied latin. So also are some of the rules of the Romans, such as all the trees being of feminine gender so that the generic name of the common beech-tree is *Fagus* (a feminine noun referring to all beech-trees), and its binomial is *Fagus sylvatica*, the specific name ending in 'a' showing that it is a feminine adjective describing exactly what kind of beech-tree is being mentioned. Most botanists must learn rather than understand all this. However, it is worthwhile being correct if one is really interested.

In horticulture special races of plants have been raised having features far removed from those of their wild ancestors but fitting them to the requirements of gardeners. These are called 'cultivars' (short for cultivated varieties) and the names are often fanciful and written thus: *Linaria* 'Fairy Bouquet' or *Gypsophilia* 'Bristol Fairy' with the species name omitted. As one becomes more and more familiar with these plants they are referred to among gardeners just as 'Fairy Bouquet' and 'Bristol Fairy'.

So important are plant names in science, in agriculture, fruit culture, the flower trade, the study of diseases (and remedies) and many other spheres that many an international conference has discussed knotty points and problems and tried to iron out difficulties which have arisen over the hundreds of years that plants have been studied. The *International Code of Botanical Nomenclature*, to-

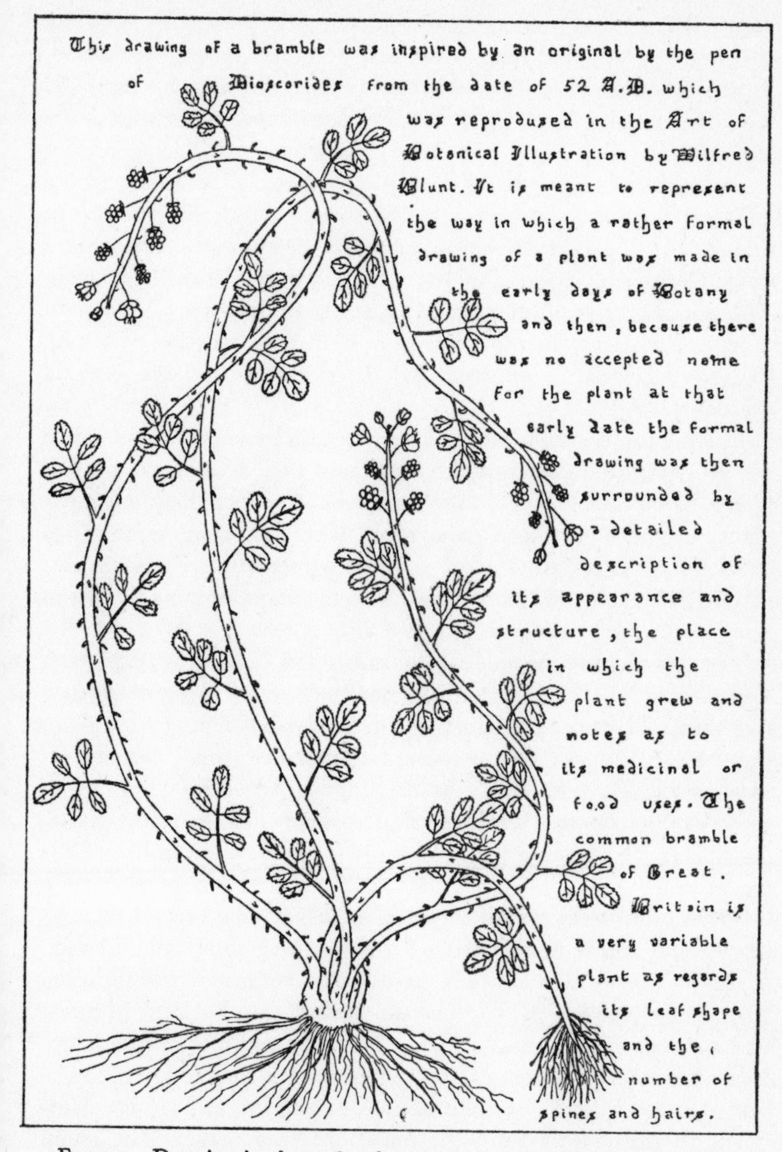

Fig. 50. Drawing in the style of an early botanical description of a plant

gether with the *International Code of Nomenclature for Cultivated Plants* which was first published in 1961 and revised in 1969 lay down recommendations which are now accepted almost everywhere. It will probably be a long time before everyone changes their old habits of referring to plants by familiar names rather than the newly revised ones.

Changing scientific names is always a source of confusion to the ordinary person, indeed even to the scientist. Basically the rule is that the oldest acceptable name is used. The confusion arises because some plants may originally have been wrongly identified, given different names by different collectors or their names lost for several decades. Some botanists spend most of their time trying to match up fresh living plants with years-old dried remains of the original specimen (see Plate 7).

The picture is somewhat confused because some botanists tend to be 'splitters' giving each slightly different and distinct plant a different name while others are 'lumpers' who tend to put similar plants in the same species and given them the same name. Very often the only way to clear up difficult points is to grow the plants in question under identical conditions to eliminate any variations due to external influences. This is a long process.

The classification of plants into orders and families is very much a matter of scientific opinion and need not worry the non-botanist too much. There is no perfect system of classification acceptable to everybody. However, one speaks of 'leguminous plants' meaning all those possessing pod-like fruits or legumes: of 'composites' as all those having compact heads of small florets as in daisies and dandelions. This is a classification in itself. It is useful to group like plants together for many purposes. Eating apples, alpine plants, cacti, anemones and trees are all classes of plants having certain features in common and these features are used for their grouping, although in some cases this may cut across the botanist's classification. Hybrids derived by crossing two different plants, are often difficult to classify or even to name. They seem to have a foot in two camps because they have arisen as seed which had mixed parentage. The male parent, or pollen parent as it is called, might have been different in many ways from the female or seed parent. The union (cross pollination and fertilisation), may have brought forth fine strong offspring. This might happen by chance with seedlings growing amongst the parent plants or the job might have been carefully planned and hand-done by a skilled plant breeder. Crossings between closely related plants are usually fairly successful and crosses between plants in different genera are rarer and more difficult. Such hybrids are given the names of both parents (if they are known), and written for example as *Sorbopyrus* indicating *Sorbus* and *Pyrus* parentage.

After the latin name it is correct to place the name of the 'author', the botanist who first described the plant and published details about it and to whose works reference may be made as in *Fritillaria imperialis*, L. showing that Crown Imperial (common name), was first described by Linnaeus. As there are, of course, many such authors, initials and abbreviations are used. In fact, the better known the author, the more his name can be abbreviated. Where plant names have been revised there may be two or more authors names mentioned.

The most comprehensive list of plant names is the ever growing *Index Kewensis Plantarum Phanerogammarum* which began its publication at the Royal Botanic Gardens, Kew in 1893 financed by a gift from Charles Darwin. A valuable recent and more easily obtainable publication on the subject is *Willis' Flowering Plants and Ferns* revised by Airy Shaw and published in 1968. The non-botanist would find both of these difficult to master.

Books which deal with plant names, classifications and descriptions are called floras. There is no single world flora, but in botanical libraries one may consult the floras of almost any part of the world and of any climate or ecological condition. (Sometimes the word flora is used simply to mean the plants which grow in a particular area). Many manuals and floras contain keys for the identification of plants. These keys demand various levels of botanical skill and knowledge but fortunately many are well illustrated, a great help to those whose botanical vocabulary is limited or halting. Most keys ask questions of the 'either/or' nature and lead logically on by stages until the plant is eventually identified. Some botanists are now working with 'computer taxonomy' which enables them to match dozens of characters at once. Ferns, fungi, lichens, conifers and micro-organisms are all classified in a similar way to the flowering plants, except that for some, of course, there are no cultivars.

The naming of plants can be the most absorbing, fascinating, infuriating and rewarding pastime or lifes' work. Whatever it is, it is a useful habit for a flower arranger to acquire. This is particularly so at the present time. People have now become increasingly precise in their demands for plants for special arrangements.

Those in search of 'grey-leaved' or 'green-flowered' subjects must know what to ask for. It has often been said that flower arrangers must 'know their onions'.

As one becomes more interested and more knowledgeable, special forms of familiar species enter into ones vocabulary. Now it may be the 'Galaxy' sweetpeas, Aster 'Cut and come again', 'Butterfly' antirrhinums, Long-spurred hybrid columbines or the ornamental grasses which are in demand. Their origins and correct nomenclature are of interest to both the botanist and the flower arranger.

Flowers which have been specially treated by growing them at particular temperatures or in controlled periods of light and darkness and which respond to these conditions by flowering out of season, often have a tag on their names denoting this fact as in 'pretreated early flowering hyacinths'. This difference is only due to the conditions of growth and is not permanent, and such bulbs would need to be treated each year to produce these effects.

Flowers which are permanently different from their near relatives are genetically separate forms and have often been produced by plant breeders and have been selected because they show attributes which make them desirable for the cut flower trade or for growing in ones garden. Such breeding usually first involves the choosing of parents showing interesting and promising characteristics such as shape, colour, size of flower or habit of growth. Then the cross-pollination of the parents must be arranged. This is the placing of the ripe pollen from one parent onto the receptive stigma of the other parent and the protection of the flower from the unwanted insect visitors which might contaminate the cross with pollen from other flowers. Simple but scrupulously clean instruments are needed for this work.

Then follows the waiting time for the production and ripening of seed and the sowing and growing of this seed and a further period of waiting for the tiny seedlings to form leaves and flowers. Often, after all this care and attention, these plants are disappointing and may need further crossing to their parents or amongst themselves until the desirable traits are produced and fixed in a new race of plants. These must be tested carefully for several years before they can be put on the market (see Plate 16).

After all this effort, it is good to know that the rights of the plant

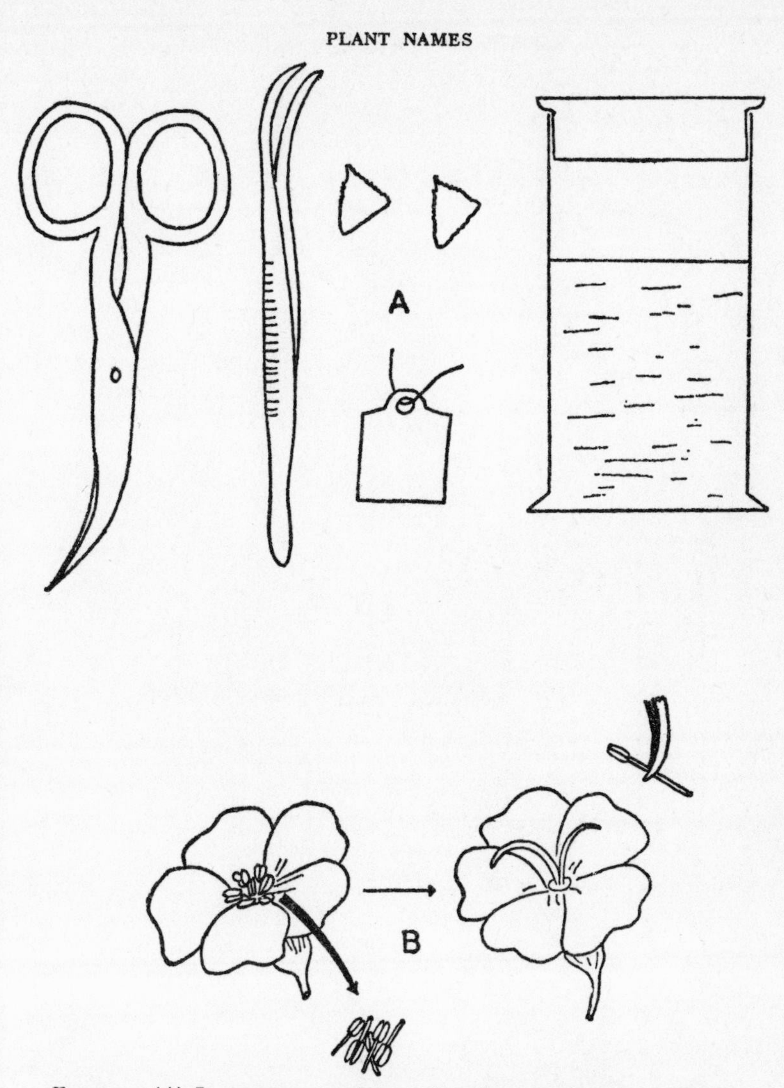

FIG. 51. (A) Instruments used in pollinating experiments. Pointed scissors used to cut off unwanted stamens; pointed forceps for transporting stamens from one flower to another; tie-on labels to record details of the cross; torn bits of blotting paper for brushing pollen out of stamens; jar of methylated spirits for sterilising all instruments. (B) A flower ready for cross pollination. The stamens have ripened first and are cut off. When the stigmas are ripe a stamen from the chosen pollen parent is brought to the seed parent and the pollen deposited on the stigma. The protective bags and record books are not shown

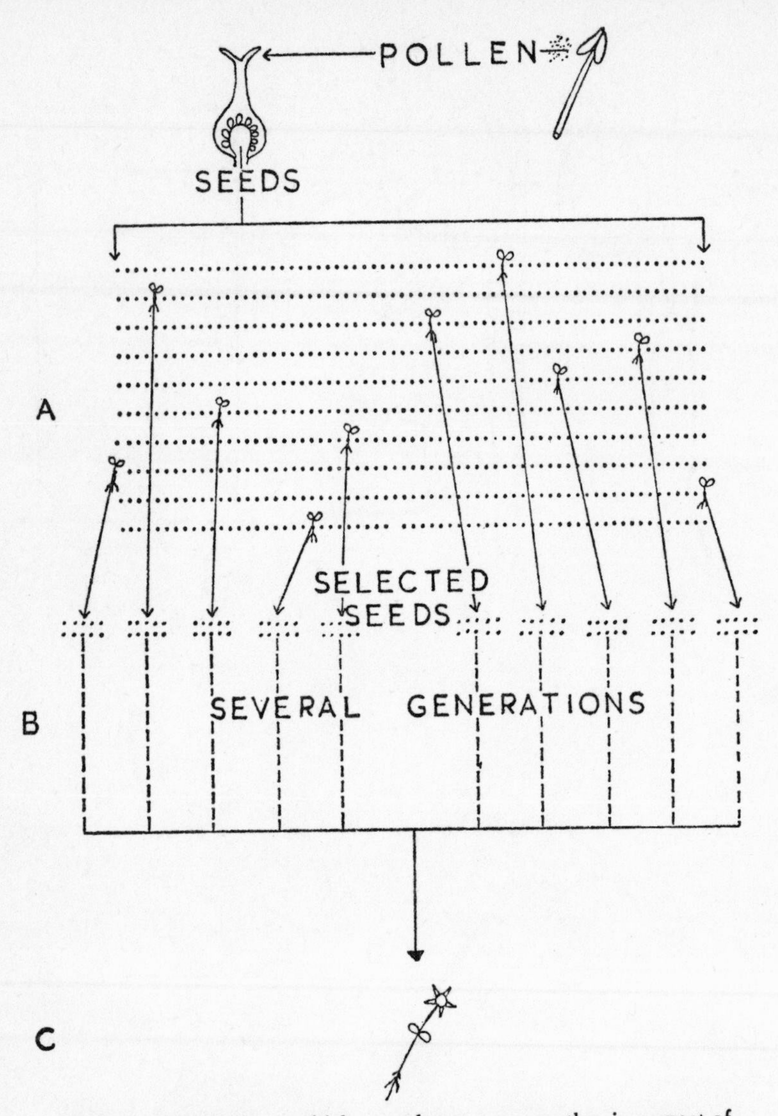

Fig. 52. Parent plants which complement one another in respect of desirable characters such as colour and scent are first selected and crossed. (A) Hundreds of seeds from this cross are then sown. (B) Perhaps 10 plants are selected because they show good features of colour and of scent. These are self-pollinated, each producing many seeds. This is repeated or several generations. (C) Perhaps one plant eventually proves desirable. It is self-pollinated and all the seed grown for trial. Only the best plants are kept for seeding each year. By continued selection the breeder is eventually a satisfied and a stock is built up for commercial trials. If succcessful a name may be given to the new race of plants which brings together the desirable traits of the two parents.

breeder have now been safeguarded by the Plant Varieties and Seeds Act of 1964 which shows that their good work has been acknowledged in the horticultural world. George Russell disliked the thinly flowered stems and restricted range of colours in lupins available in his day and worked for many years selecting for better size, colour and arrangement of flowers and died at the age of 94 in 1951 without the protection afforded by this Act. The results of his work may be seen in the vast range of Russell lupins available for everyone to enjoy. Among many gratifying rewards for his work, however, he gained the much coveted Royal Horticultural Society Gold Medal and the Veitch Memorial Medal and, in the year of his death, the M.B.E.

Protection for discoverers and originators of varieties in the U.S.A. is provided by the Trade Mark and Patent Laws. Under the Federal Seed Act of 1939 (amended in 1956), the originator of a new variety has the right to name that variety, but when a trade mark is made a name of a variety, it is no longer protected as a trade mark. Details specify that the applicant for a patent of a particular variety must have done something to create or produce it and it must be distinct and new. A plant found by a person is not considered patentable.

Occasionally a new plant of good and desirable features suddenly springs out of the blue in a garden as did the 'Spencer' sweet pea in 1900, the first to have the waved petals so familiar in modern races of sweet peas. This sudden change in the genes controlling inheritance in plants is called mutation and the resulting plant a mutant or sport. Gardeners and nurserymen are always on the look-out for this to happen. The rose 'Kronenbourg' grown for its beautiful red colouring arose as a sport of the well-known 'Peace'. A tiny white-leaved, crimped sport 'Watsoniana' arose from *Rosa multiflora*, the seeds produce only *R. multiflora* plants while cuttings will perpetuate 'Watsoniana'.

The diagram which ends this chapter represents the possible origin of a new variety.

FURTHER READING

Flowers and their Histories by A. M. Coats (Hulton).
Manual of Cultivated Trees and Shrubs by A. Rehder (Macmillan).
Manual of Cultivated Plants by L. H. Bailey (Macmillan).

Identification of Trees and Shrubs by F. K. Makins (Dent).
Herbaceous Garden Flora by F. K. Makins (Dent).
Know your Conifers by H. L. Edlin (H.M.S.O.).
Garden Flowers by R. D. Meikle (Eyre & Spottiswoode).

CHAPTER EIGHT

Some More Botany

Flowers

Flowers are special shoots modified for reproduction. They are made up of whorls or rings of special floral parts some, such as sepals and petals being sterile and appearing to perform functions of protection of the flower when in bud and attraction of insects for pollination. The single ring of petals may be multiplied in 'double' flowers which consequently look more colourful. Other floral parts are associated with sexual reproduction; stamens producing pollen in which male gametes develop and carpels producig ovules in which ova or egg cells develop. Most flowers are hermaphrodite (bisexual), producing both male and female gametes and comparatively few are unisexual, either male (staminate) or female (pistilate).

Sepals are usually green and are grouped into a calyx. In plants such as anemone they are brightly coloured and said to be petaloid. The corolla of petals is normally coloured. Much work has been done on the colour appreciation of insects which are the principal pollinating agents. Many insects can appreciate colour, bees, for instance, being attracted to yellow, blue-green and blue flowers (red probably appearing the same as blue-green to bees). Honey guides, markings which indicate the way towards the nectaries, are lines of contrasting colour situated on the petals and may even be invisible to humans and visible to insects. The shape of the flower and its scent are also appreciated and 'remembered' by insects encouraging them to be constant visitors. This is important to flowers because they can only utilise pollen of similar plants.

Those flowers which do not produce attractive sepals or petals may be wind pollinated (see note about catkins).

Whereas the calyx and corolla form a floral envelope and are regarded as accessory parts of the flower, the stamens and carpels are considered as essential parts.

Stamens are made up of a stalk or filament which carries a small vein from the base of the flower to the expanded tip, the anther, wherein pollen develops in pollen sacs. Pollen is shed when the

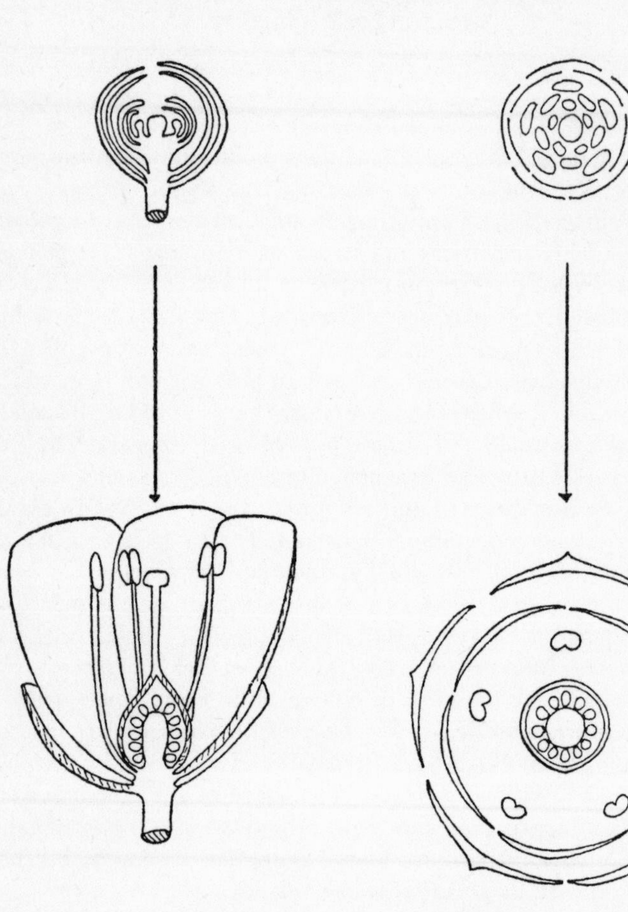

FIG. 53. At first most flower buds look similar, the separate floral parts at this time are tiny primordia made up of growing and differentiating cells. These primordia develop into the characteristic floral parts, sepals, petals, stamens and ovaries.

anthers open on ripeness. Pollen is often yellow and powdery but is highly individual, the pollen of different species being recognisably different when seen under a microscope. Pollen is a nutritious food for young insects and the act of pollinating flowers is performed during the chore of food collecting.

The female part of the flower is the carpel, one or more forming an ovary above which is a narrow style terminating in a receptive stigma on which pollen must be deposited before fertilisation can take place. This occurs when a pollen tube grows down the style, stimulated by hormones, and carrying the male gamete right to the egg cell where fusion takes place. The egg cell within the ovary then develops into a tiny embryo enclosed in a seed coat and protected by the ovary wall which forms the fruit.

Great significance is placed by botanists on the position of the ovary with respect to other parts of the flower. If the ovary is attached to the floral receptacle above the insertion of the petals it is said to be superior; if it is placed below the insertion of the petals is said to be inferior. The superior ovary of the flower of antirrhinum is enclosed by the petals and the fruit is only obvious when the petals have fallen. The inferior ovary of a daffodil can clearly be seen below the flower and the capsular fruit is at first topped with the remains of the dying flower.

Nectaries are often situated at strategic places causing insects to pass stigmas and stamens on their way in and out of the flower.

Flower Shapes

A. A regular (actimororphic), flower may be cut into equal halves along any radius. Such flowers may be open or tubular. Open flowers (buttercup and rose), are accessible to all sorts of insects and, as their nectar is not deep seated, they are visited by short-tongued insects. Tubular flowers may be made up of petals joined into a funnel-like tube or of free petals shaped to form a tube at the base of which are the nectaries. Often such flowers are also shaped to have a flat area which can provide a landing stage for visiting insects. These flowers (primroses and wallflowers), are pollinated by long-tongued insects for only they can reach the nectar.

In pollination it is essential for the insect to come in contact first with the stigma (which collects pollen from previous flowers), and then with the stamens (which deposit the next load of pollen). In many flowers the stigmas and stamens do not become ripe at the

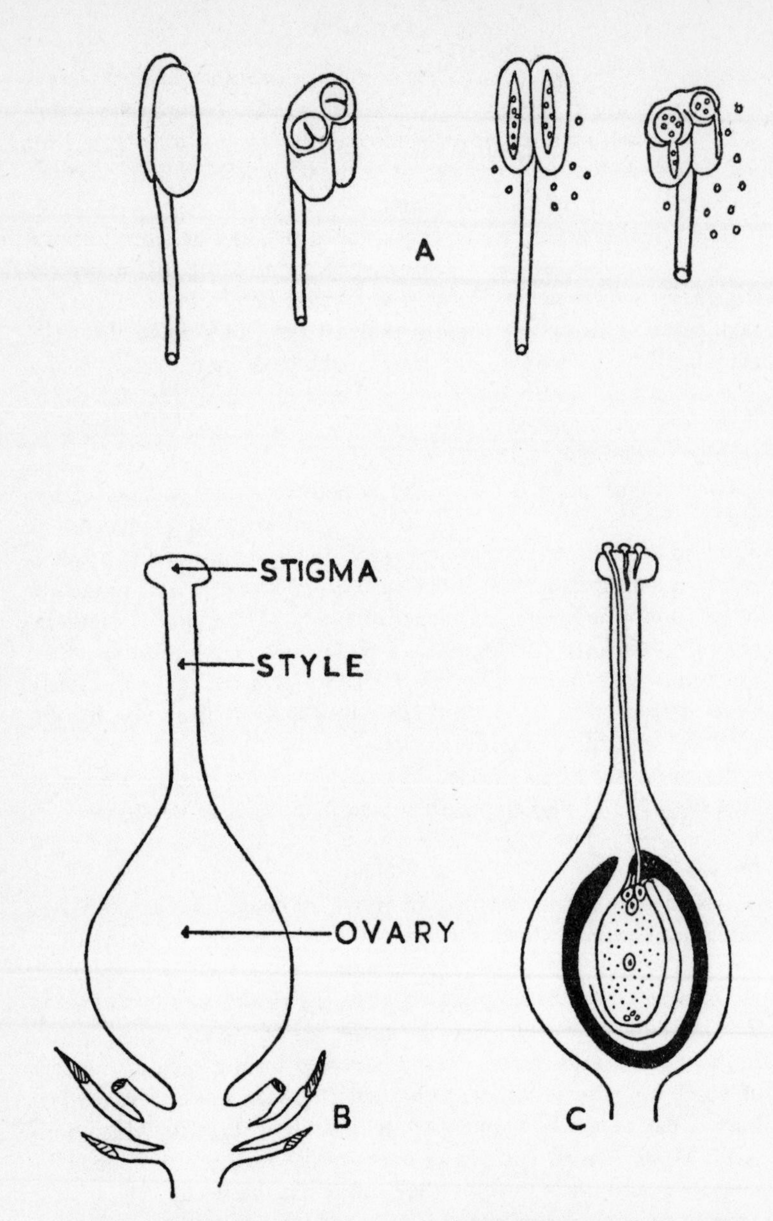

FIG. 54. (A) Views of stamens to show the pollen sacs before and after splitting to shed the pollen. (B) General view of the pistil (female parts of the flower). (C) Section of the pistil showing pollen tubes growing down the style, one in contact with the egg cell of the ovule

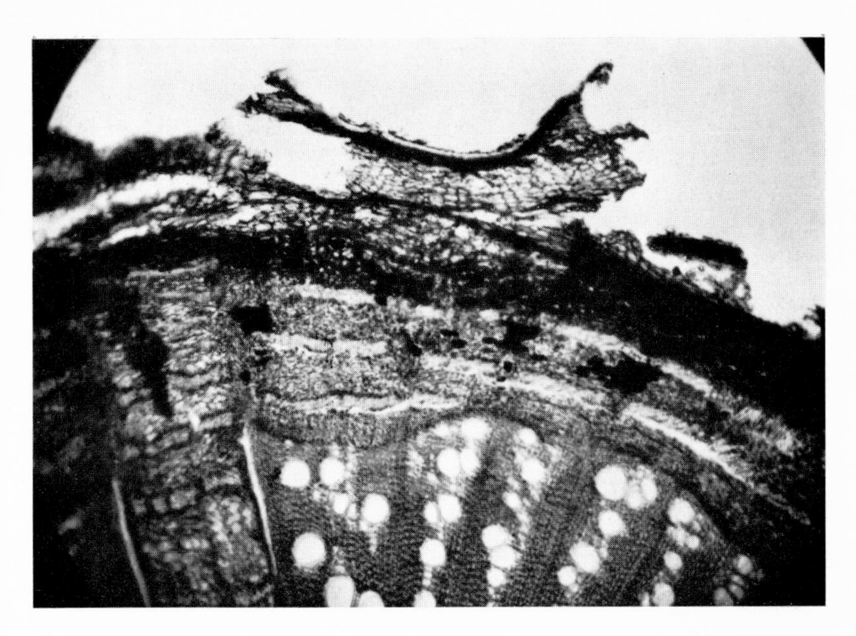

Plate 13. Detail of a section of a woody stem showing the bark flaking off.

Plate 14. Unusual pieces of wood with interesting shapes which can be of use in displays.

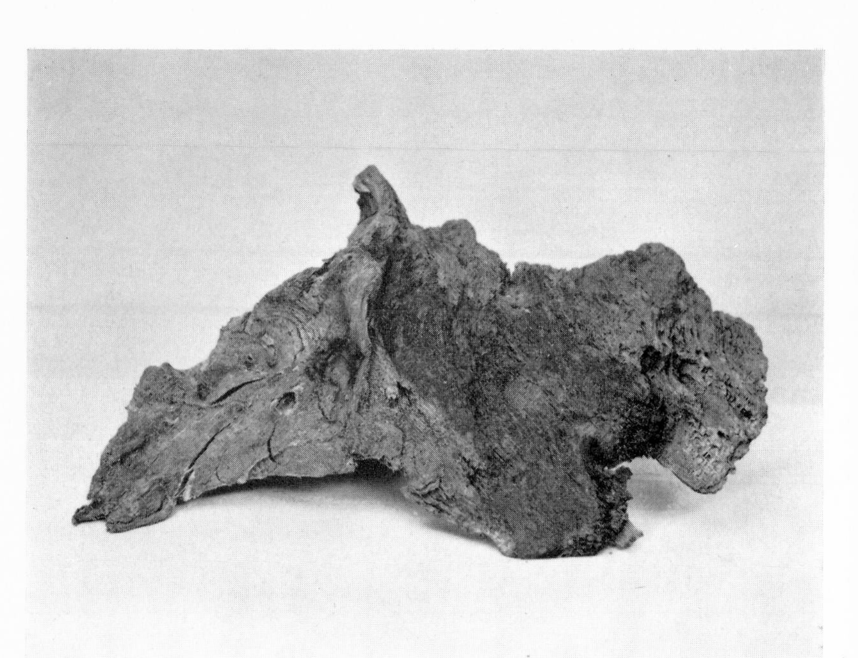

Plate 15. Hard, heavy wood from the root of the Spanish Chestnut tree. (Lent by Trevor Sears of Wormley, Surrey and photographed by Muddle of Godalming.)

Plate 16. One of the R.H.S. Flower Trials at their Wisley gardens. Expert judging may result in the Award of Merit for a cultivar.

same time and such flowers must encourage insects to visit them at least twice.

B. An irregular (zygomorphic), flower is one which cannot be cut into equal halves along more than one radius. Such flowers offer shapes which seem to be attractive to insect visitors, give a good landing place and, by secreting the nectar deeply, perhaps at the end of a long spur, a reward to only long-tongued insects for their part in pollination. It is of interest to note that specialisation in flower structure leads to 'flower constancy' in insects and instead of any insect visiting any sort of flower, insects and flowers seem closely allied. In this way less pollen is wasted and the chances of suitable pollination is increased. Pollen is known to be highly specific and it will only function on stigmas of the same species or variety.

C. A capitulum or head of small flowers or florets is seen in members of the daisy family. This is an inflorescence in which the tight grouping of many small flowers is possibly a form of economy, all massing together for attraction, concentration of scent and to provide a landing stage for insect visitors. The whole capitulum is protected in the bud stage by a number of bracts (called an involucre), which act rather as sepals in other flowers, closing them and protecting them when in bud.

Capitula may be composed entirely of flowers of similar appearance as in a 'double' chrysanthemum and a dandelion or of two different types; ray florets around the outside and disc florets forming a centre, as in the sunflower (*Helianthus sp*). Usually the ray florets are female, bearing a pistil but no stamens, while the disc florets are hermaphrodite (also called bisexual or perfect), bearing stamens as well as a pistil. Cornflowers (*Centaurea cyañus*), and knapweeds (*Centaurea nigra*), have a central cluster of disc florets surrounded by a ring of funnel-shaped sterile florets whose only function is to make the flower head more attractive and easily seen by insects.

When the florets set seed the tiny one-seeded fruits may be plumed and feathery and are dispersed by the wind.

Catkins and Cones

Catkins are long, clustered, usually pendulous inflorescences made up of small flowers which are wind pollinated and are characteristic

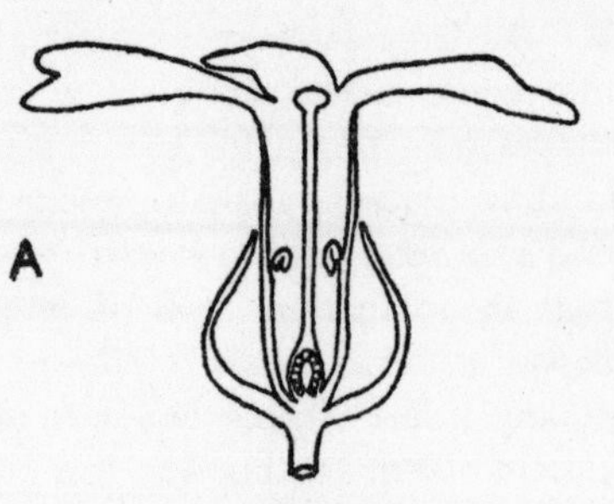

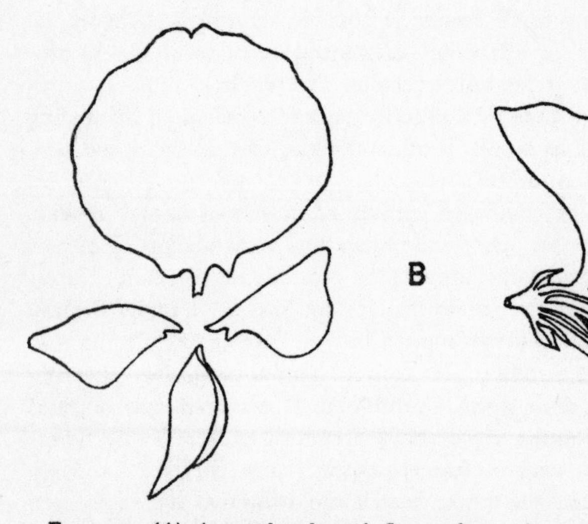

Fig. 55. (A) A regular-shaped flower from above showing overlapping petals, and the same cut lengthways to show the internal parts. (B) An irregular-shaped flower of the pea family showing the separate petals, the standard, the two wings and the folded keel, also cut lengthways to show the internal parts

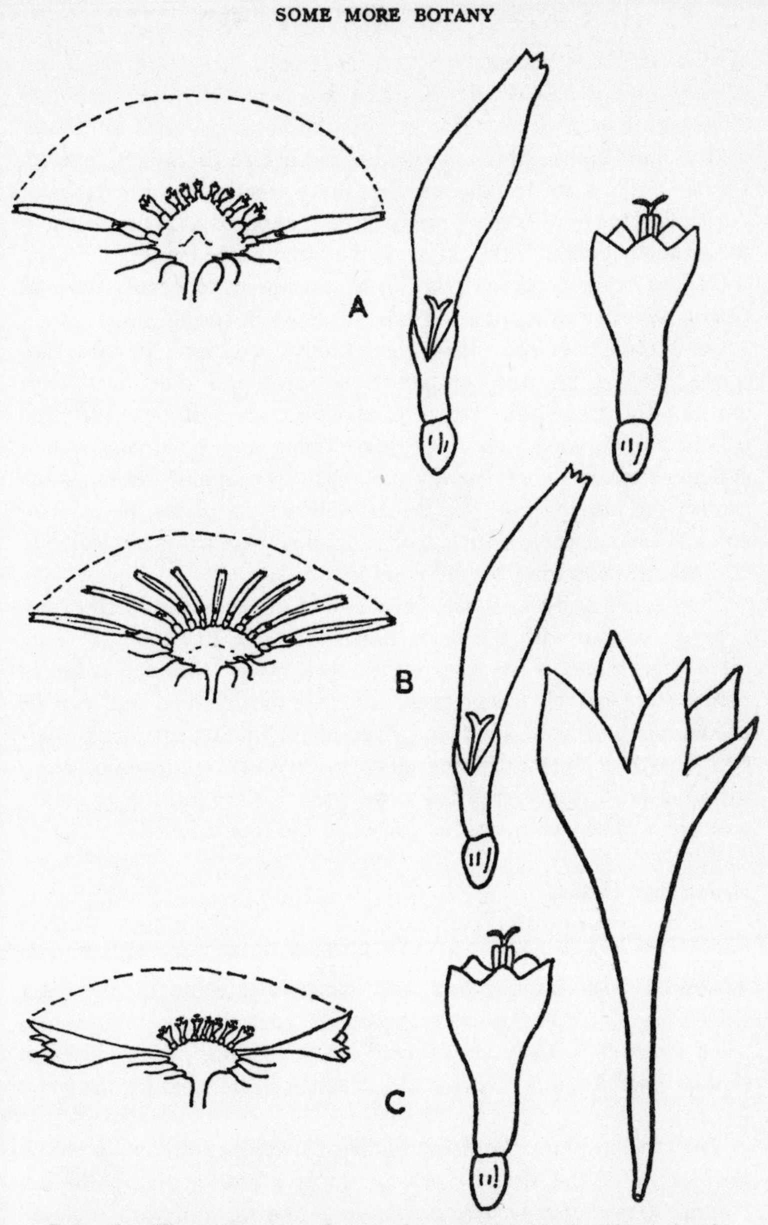

FIG. 56. The component florets of composite heads. (A) A head composed of ray and disc florets. (B) A head composed entirely of ligulate, strap-shaped florets. (C) A head composed of fertile disc florets surrounded by larger, attractive florets which are sterile

of many trees such as species of oak (*Quercus*), hazel (*Corylus*), birch (*Betula*) and willow (*Salix*). Catkins are unisexual and sometimes the sexes are situated on different trees. Male catkins produce masses of dry, light pollen which is usually shed before the leaves develop. Female catkins are usually smaller, more erect and the stigmas of the individual flowers are prominent and exposed in order to catch the airborne pollen. They often develop into nutty fruits.

Catkins make good material for arrangements especially those of *Garrya elliptica* in winter and pussy willows in spring.

Coniferous trees bear their reproductive structures in unisexual cones. Male cones are comparatively small and short-lived with overlapping stamen-like scales producing masses of light airborne pollen. Female cones are slowly developing and eventually woody structures made up of spirally arranged, overlapping scales which protect the naked seeds. (Unlike other flowering plants the seeds of conifers are not enclosed in an ovary and no true fruits are formed). The seeds are carried by the wind.

Tiny cones of cypress-trees (species of *Cupressus* and of *Chamaecyparis*) contrast with the large cones of the true fir (*Abies procera*), more than 6 inches long, and the erect, heavy, rounded cones of cedars (*Cedrus sp*). Many cones are periodically shed and can be picked up and used over and over again in arrangements, their light-brown colouring lending much to very varied themes of decoration; others, firmly attached to the branches are difficult to obtain and are a prize worth waiting for when the tree is felled.

Arums and Orchids

Aroids, or arum-like plants belong to a large, mainly tropical family of monocotyledonous plants distinguished by tiny flowers associated with large bracts and comprising some of the most decorative plants in flower arrangements. Basically the inflorescence spike (spadix) is variously covered by small unisexual or bisexual flowers and the main feature of attraction is afforded by the bract (spathe).

The cuckoo pint or lords and ladies (*Arum maculatum*), a woodland plant of the British Isles, produces a pale green spathe encircling a deep purple spadix, sterile at the tip and with separate male and female flowers, ripening to form red berries, a popular subject for autumn arrangements.

The white arum or lily-of-the-Nile (*Zantedeschia aethiopica*),

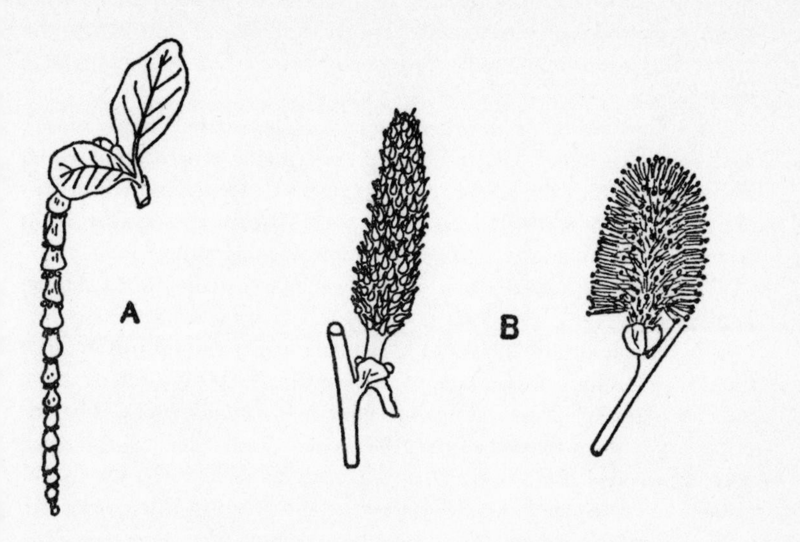

C

FIG. 57. (A) A male catkin of *Garrya elliptica*. (B) Female and male willow catkins. (C) Views of female pine cones showing woody scales, both open (dry) and closed (damp). Also a single scale bearing winged seeds favoured for church and formal decorations is made up of an erect spadix covered with tiny female flowers in the lower part and in the upper part with yellow male flowers and ensheathed by a large, pure white bract.

Flamingo plant or painters palette (*Anthurium andreanum*), with small, clustered bisexual flowers covering an erect spadix which stands away from the leathery scarlet or pink spathe, and white sails (*Spathiphyllum wallsii*), with small erect spadix and spathe seem to impart elegance and extravagance to arrangements.

The leaves of aroids, some of the best house plants, are valuable in their own right with long petioles, and beautiful shapes.

The complexity of the floral structure and its variety in the large family of orchids makes description difficult. Most orchids have stalkless (sessile), flowers forming spike-like inflorescences. The inferior ovaries are twisted so that the flowers often 'face' all the same way. The waxy, long-lasting tropical orchids need to be seen (and grown) to be believed. Ladies slipper orchid (*Cypripedium sp* or the related *Paphiopedilum sp*), are solitary flowers, the outer perianth being coloured and the central sepal broad while the two laterals are very small. The inner perianth or petals are remarkable for the too thin, often twisted laterals and the large slipper-like labellum. The seeds of orchids are very small and need contact with a fungus for germination to take place. The seedlings grow slowly to form perennial plants with swollen bases of pseudobulbs. Although many orchids are terrestrial some are epiphytes growing on tree branches and having aerial roots.

The Inflorescence

The arrangement of flowers on a plant is known as the inflorescence. A flower may be solitary at the end of a stalk (called a scape if it arises directly from the ground), or grouped in various ways which are constant and therefore typical and peculiar to a plant. In general inflorescences are of two types. In one (racemose), the oldest flower is at the base and there are progressively younger ones towards the tip. In the other (cymose), the oldest flower is terminal and the younger flowers appear lower down the stalk. There are many forms of racemose and of cymose inflorescences which can be recognised in decorative plants. Some are illustrated here.

Leafy bracts are often associated with flowers and in umbels and capitula are clustered together to form an involucre.

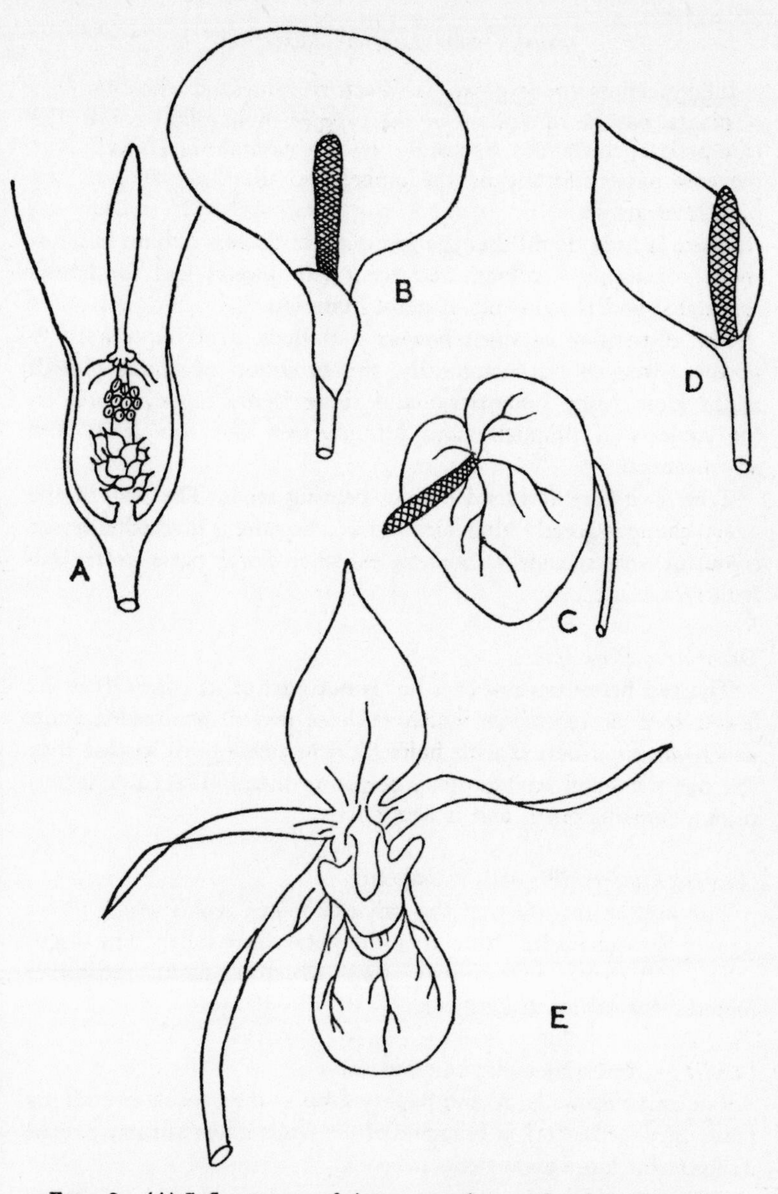

Fig. 58. (A) Inflorescence of Arum maculatum, showing the sterile tip of the spadix below which a ring of hairs closes the entrance to the upper, unisexual male (staminate) flowers and the lower, female (pistilate) flowers. Female flowers ripen first, thus the visiting insect deposits pollen on the stigmas before the male flowers shed their pollen. The hairs then wither and the insect escapes to another arum. (B, C and D) Other aroids mentioned in the text. (D) Ladies slipper orchid

Inflorescences are of great classificatory value and whole families of plants may be recognised by the type of their inflorescence: the Compositae, the sunflower family, by the capitulum, Umbelliferae the cow parsley family by the umbel and the Gramineae by the panicle of grasses.

There is little doubt that the grouping of flowers into an inflorescence concentrates colour, and scent (for insects and for flower arrangers) and is economic of plant materials.

The disposition of open flowers and buds is of importance to flower arrangers but botanically, the evolution of flower growth is far from being understood and there is much speculation on the subjects of simplicity and primitiveness and complexity and advancement.

True fruits are fertilised ovaries bearing seeds. The wall of the ovary changes greatly after fertilisation, becoming markedly dry or colourful and succulent. Sometimes other floral parts enter into fruit formation.

Decorative Rose Hips

The red fleshy portion of a hip is not part of an ovary. It is the flower base or receptacle which encloses several one-seeded fruits which are interspersed with hairs (said to tickle birds so that they spit out the small fruits instead of eating them). The calyx is persistent, remains green and is attractive.

Chinese Lantern (Physalis franchettii)

The lantern is formed of the calyx of joined sepals which grows around the tomato-like fruit when the petals have fallen. The bright colour and papery texture make these structures useful everlasting material for winter decorations.

Shoo Fly Plant (Nicandra physalodes)

The calyx grows large and papery around the capsule containing many tiny seeds. Forked branches of the fruits make attractive dried material for large arrangements.

Leaf Types

Leaves are the most varied of all the vegetative parts of the plant. Typically a leaf is composed of a petiole or stalk and a lamina or leaf blade and a pair of expanded outgrowths at the base of the

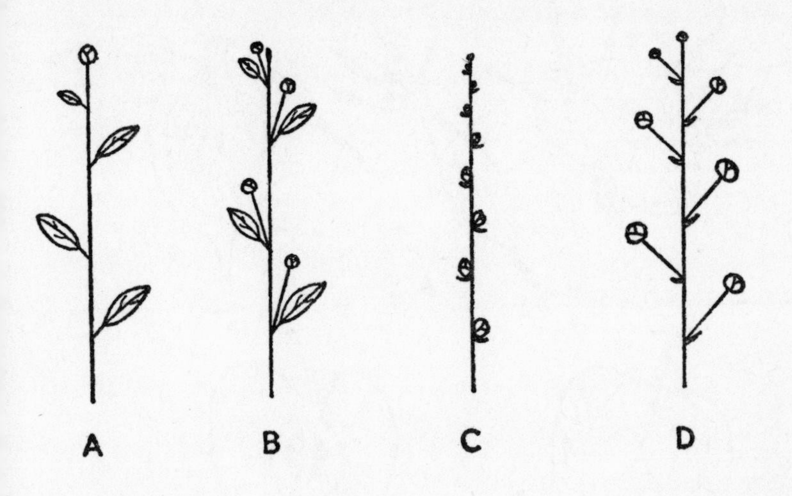

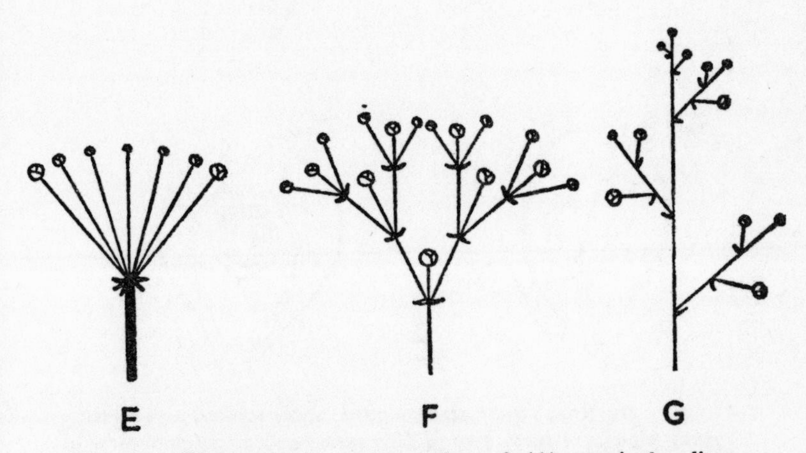

FIG. 59. Diagrammatic representation of (A) terminal solitary flower, (B) axillary solitary flowers, (C) spike, (D) simple raceme, (E) umbel, (F) cyme, (G) panicle. The capitulum is shown in Fig. 56

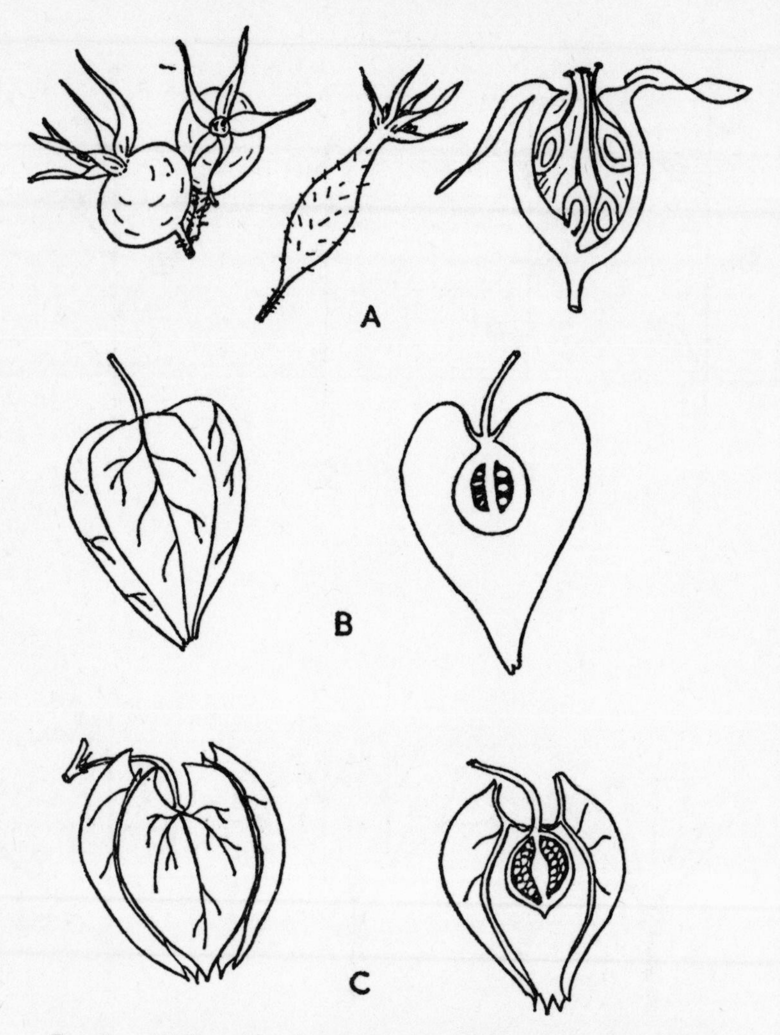

FIG. 60. (A) Rose hips. Large rounded, apple-scented hips of *Rosa rugosa*, elongated spiny hips of *R moyesii*, and an internal view of a hip showing the single-seeded fruits attached to the succulent receptacle. (B) Orange coloured papery Chinese lantern showing the small tomato-like fruit completely enveloped by the enlarged calyx. (C) Dried sepals enclosing the capsule containing many seeds in the Shoo fly plant

petiole, the stipules. Leaves without stalks are said to be sessile, and leaves without stipules, exstipulate.

The arrangement of the vascular bundles or veins is called the venation. Parallel venation (seen in many monocotyledonous plants) shows several principal veins extending from the base to tip of the leaf, whereas net venation shows a prominent midrib and smaller lateral veins forming a fine network.

Leaves may be simple, having but one blade, or divided into lobes, or completely divided into individual parts or leaflets. If the lobes or leaflets are arranged on either side of the main vein the leaf is said to be pinnately divided or pinnately compound. If the lobes or leaflets diverge from a common point at the end of the petiole the leaf is said to be palmately divided or palmately compound.

Leaves which fall at the end of a growing season are said to be deciduous; those which remain on the plant for more than one season are said to be persistent and the plant, evergreen.

Leaf Shapes

Leaf shapes are very variable and botanists have given them a great many names which are largely descriptive and self-explanatory.

It is usual for the leaves of a plant, though they may be variable in size due to age or colour, and in texture due to exposure to light and shade, to be all of roughly the same shape. In some plants, however, basal leaves forming a rosette may be quite different in shape from those situated on erect parts of the stem. In other plants leaves near flowering shoots may be different in shape from those on completely vegetative parts.

Perhaps the shapes of leaves and their arrangement on the stems and branches are related because in many cases distinct leaf-mosaics can be seen where leaves fit up against each other without much overlap, almost like pieces of a jig-saw puzzle.

The surface of leaves may be smooth (glabrous), or covered with hairs, scales or various kinds of coatings. Hairs may be stiff, silky or even woolly and matted (tomentase) and often impart a grey colouring to the leaf. This may be of great help to flower arrangers who can use this gentle colouring as a foil for other brighter or subtle shades. Grey-leaved plants are often found growing naturally in areas exposed to sun and wind. The hairy covering may act as protection against excessive water loss by these plants.

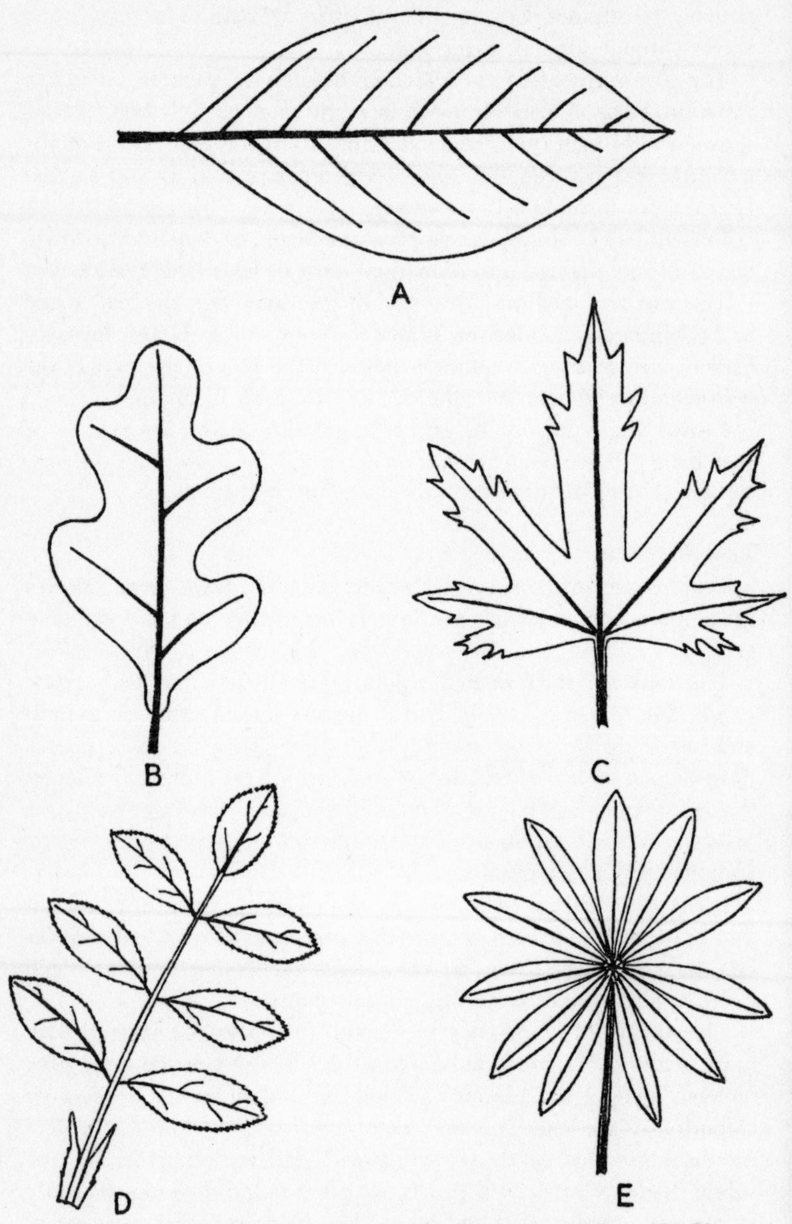

FIG. 61. Leaf types. (A) Simple, (B) Pinnately divided, (C) Pinnately compound, (D) Palmately divided, (E) Palmately compound

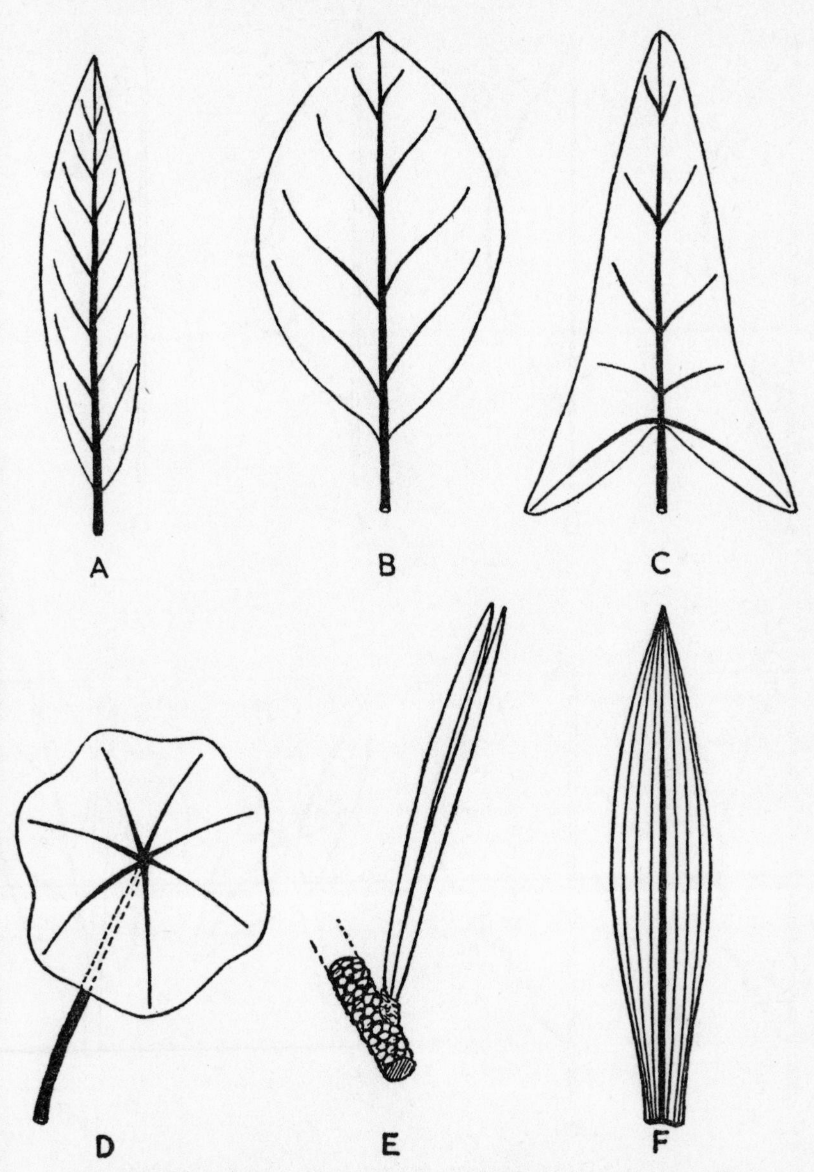

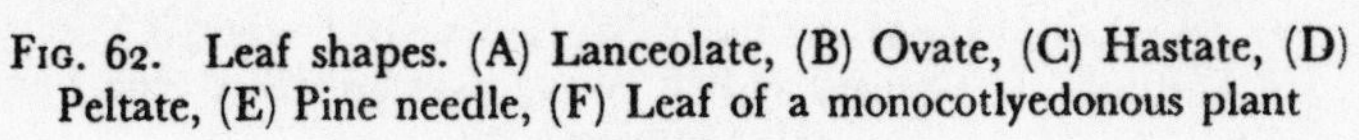
Fig. 62. Leaf shapes. (A) Lanceolate, (B) Ovate, (C) Hastate, (D) Peltate, (E) Pine needle, (F) Leaf of a monocotlyedonous plant

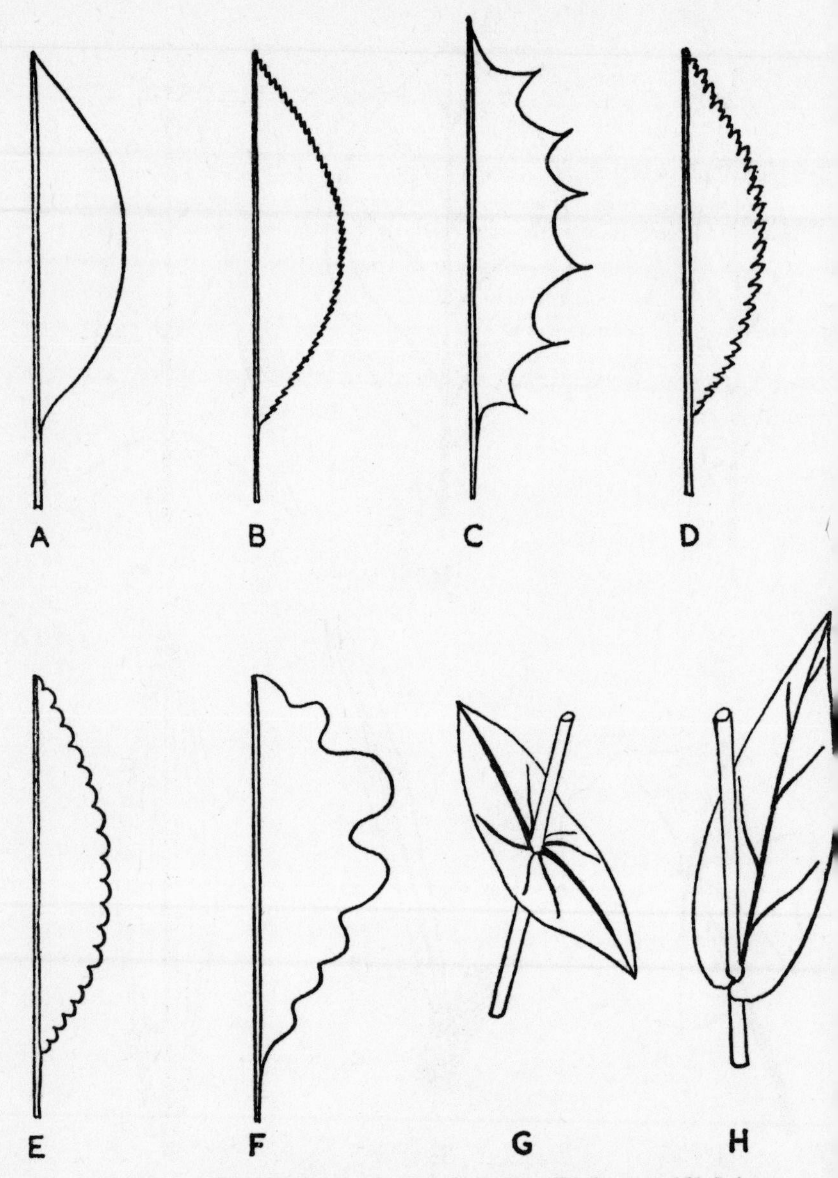

FIG. 63. Leaf margins and bases. (A) Entire, (B) Serrate, (C) Spiny, (D) Dentate, (E) Crenate and (F) Wavy margins, (G) Connate and (H) Clasping bases

A glaucous leaf is one covered by a blue or whitish 'bloom' which can be removed if the leaf is handled. Glaucous varieties of plants are in great demand for arrangements and impart soft, pleasing shades.

Leaf Margins and Bases

Leaves show variety at their bases, their tips and their margins. Careful attention to these details makes descriptions of leaves accurate and useful in plant identification. Often details about the leaves of at plant are of such importance that that feature is incorporated in the plants name. This is so in *Fagus sylvatica* 'Laciniata' a cut leaved variety of the common beech; in *Campanula persicifolia* the peach-leaved bellflower and *Hedera helix* 'Sagittaefolia' a form of ivy having arrow shaped leaves.

FURTHER READING

The A.B.C. of Plant Terms by J. G. Cook (Merrow).
Dictionary of Biological Terms by M. Abercrombie (C. J. Hickman and M. L. Johnson (Penguin).
Dictionary of Garden Plants by R. Hay and P. Singe (Michael Joseph and Ebury Press).
The Biology of Flowers by W. James and A. Clapham (Clarendon).

Index